THE GOD POINT

THE UNDERSTANDING OF
MYSTERIOUS LIFE BETWEEN
ZERO AND INFINITY

SUNIL KUMAR

INDIA • SINGAPORE • MALAYSIA

Notion Press

Old No. 38, New No. 6
McNichols Road, Chetpet
Chennai - 600 031

First Published by Notion Press 2020
Copyright © Sunil Kumar 2020
All Rights Reserved.

ISBN 978-1-64760-733-3

This book has been published with all efforts taken to make the material error-free after the consent of the author. However, the author and the publisher do not assume and hereby disclaim any liability to any party for any loss, damage, or disruption caused by errors or omissions, whether such errors or omissions result from negligence, accident, or any other cause.

While every effort has been made to avoid any mistake or omission, this publication is being sold on the condition and understanding that neither the author nor the publishers or printers would be liable in any manner to any person by reason of any mistake or omission in this publication or for any action taken or omitted to be taken or advice rendered or accepted on the basis of this work. For any defect in printing or binding the publishers will be liable only to replace the defective copy by another copy of this work then available.

CONTENTS

Author's Words 5

1. Introduction 7
2. A Brief about Fillers of Life 10
3. Role of Symmetry 26
4. The Anti-Gravity Effect 35
5. Mysterious Life 55
6. Why 59
7. How 92
8. What 125
9. Conclusion 142

AUTHOR'S WORDS

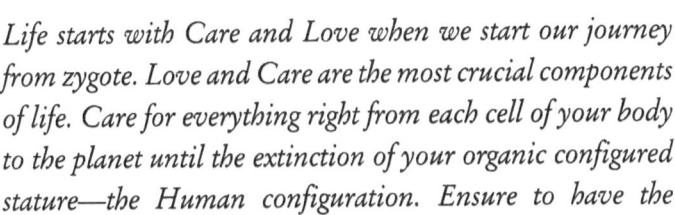

Life starts with Care and Love when we start our journey from zygote. Love and Care are the most crucial components of life. Care for everything right from each cell of your body to the planet until the extinction of your organic configured stature—the Human configuration. Ensure to have the ability to remain cent per cent happy and caring, which are the original characteristics of true Homo sapiens.

The suppliable, serviceable, consumable, transferable, bearable, unbearable, rewardable and punishable artefacts made by Humans are certainly required for us during a time-bound dwell under a civilisation but to a certain amount and coherent manner. That should not be taken to a suicidal or devastation stage.

We must be as natural as we were intended and designed to be.

There is nothing beyond this life to earn interest for the unutilised time and happiness by not utilising today's happiness, temporary possessions and relationships.

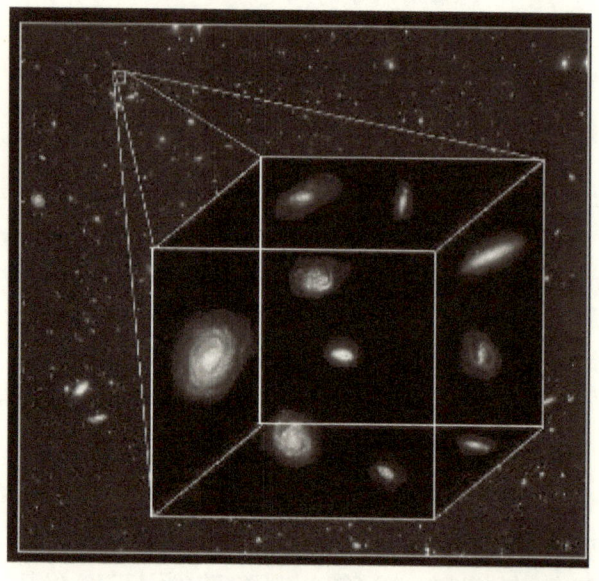

"The pre-existence indeed is the basic state of the Universe, which has to be in the state of a filler of a large giant void without any alternative choice. The Universe what we understand today is a subset of a Large Void which is eternal, infinite and partially invigorated with a scope of perpetual expansion."

INTRODUCTION

The God Point is invisible, it's a point of ecstasy, it's a ubiquitous state in the Universe, a Scientific Heaven and technically where God could possibly retain and observe "The Lives in Galaxies" and "The Earth".

The God Point

... A Concept from Zero to Infinity...

Readers to contemplate about bringing back all the planets and celestial bodies of the entire Universe into a single axis. Let's start rewinding the process of Space and Time once all celestial bodies or masses (irrespective of their physical state) are Uniaxial. We are bound to reach a point of instigation "The Big Bang Point", which is nothing but Zero Point representing coordinates to 0, 0, 0. This is the absolute "God Point". Eventually, the Universe had expanded but "The God Point" remained steady and independent of the progression of Space and Time.

Today, when an Earthian tries locating The God Point, s/he could trace it to be located at (∞, ∞, ∞) with respect to him/her. Therefore, The God Point is located at Infinity, and an astronaut can see and interpret The God Point in proximity than our planet-mates. The God Point of today is the relative "God Point" and at Infinity despite both coinciding. The God Point is where the Supreme entity can co-exist in a scientific manner.

"The God Point – The Supreme and absolute in displaced point of the Creation."

An astronaut who is trying to see The God Point is in a Proximal State of absoluteness as she is free from the Planet Myths. I mean, when we are free from Earth, gravity, and other earthy facts, we are closer to scientific reality than the Myths of Earth. Technically, an astronaut can be the ideal case to proximally interpret life better than Humans and discern truth, lie, fantasy or science. While it remains a dream for Humans on Earth to see such a point present at Infinite multi-dimensional points, where the probable Universe could have started, an astronaut who is in between Earth and God, can likely see God Point looking into proximity, theoretically. The God Point and an astronaut will unravel the five chapters of this book with the ideas, concepts and conjectures by being in The Reconnoitring Journey to "Mysterious Life" through the questions of How, What and Why of life. The God Point clarifies "why our life is like a gift from such a massive and monumental parent—the Universe". When an astronaut thought of the beginning of Life and the

Universe, she understood that there must be a point of origin. How to reach that point from now? comes to her mind, "The retrospection of the Big Bang", which means "The Rewind of the Universe and Life Formation". Such ideation can give us an idea of a point of germination or Zero Point or Today's Infinite Point—*The God Point*. The book is about reconnoitring the journey to the mysteries of life between Zero to Infinity. The life journey answers three essential questions of life—Why, What and How and with a concluding chapter of contemporary life. The Conclusion and The Mysterious Life are significantly important for 2020 Homo sapiens who must accentuate right practices in their lifestyle and interpret the absolute sense of life. To interpret unbiasedly, an astronaut who is free from gravity is considered to understand God Point. Chapter 1 is about events fillers and Life Entropy followed by the chapters Symmetry, Gravity and Mysteries Life. The conclusion is a must-read. The whole idea of 'mordern life' of Humans and 'how to live a wonderful life—A View of God Point.'

A BRIEF ABOUT FILLERS OF LIFE

Life, which passes through numerous mini battles, finally ends up in a major battle with the self and soul. The moment of finite separation can be alluded as the Detachment Moment as we have no option to stay back and dwell on the Earth. Everyone is aware of an end that might come like a bolt from the blue, an eternal sleep as right as rain. This ultima of a life book is a blistering rewind of eight different feelings felt concurrently, which fade away with subtleness. The thraldoms start becoming more delicate than they were and finally disappear on the spur of the moment. The entity, which releases from confinement, is directionless because of liberation and randomness present everywhere.

The future will never repeat the present, as the present never repeats the past. This is because of the spread of randomness everywhere in every event. No one will ever predict the future or had predicted the past. If an experiment was ever conducted to repeat one past event by artificial means, that would never be successful when viewed at a billionth level (*level of similarity between two events when a billion entities are under consideration as event makers*). If someone cancels all noise between two events, which are significant, the cancellation is meaningful as the gap or void is nothing but a filler. A filler may be tending to prioritise when viewed from a part but when we look at the overall inclusive of two endpoints, we can eliminate filler or any modifier to filler. Filler is nothing but time travel with many linked activities and random activities, which add their parts to events as equal to the surrounding.

Let us have an illustration depicting fillers:

$$\boxed{\text{Start}} - \boxed{1\text{-}2\text{-}3\text{-}4\text{-}5} - \boxed{\text{End}}$$

We can see two important events with Start and End, and the rest 1 to 5 are fillers, which are linked to each other involving millions of uncountable activities within them affected by themselves, surroundings and both combined as well. After an explicit analysis of life in context with the depicted illustration, we can conclude that both the ends are significant, and the rest are nothing but events overlapping under the spread of randomness.

The question of time travel never comes, if travelling from the time frame of infinity (∞) position to that of the infinity+1 ($\infty+1$) position makes no difference. The possibility of the same person who can be omnipresent concurrently and can travel in both positions (∞ and $\infty+1$) is just ruled out, as both the time frames will never lose their ingenuity and virtuousness of being unique frames. The most interesting part of any religion is upkeeping of records of an individual's activities especially sins and few good deeds for appropriate action on the judgement day or any synonyms of that with respect to respective religions. How to keep such records of ultra-gigabytes? Is it through the pictures of each frame from billion frames? There is no absolute necessity to start digging for responses or uncertain answers to high potential questions at this moment and, of course, "passing time will unwind some more mysteries, generating further questions and answers."

People often worry beyond their consciousness and coherency when things get out of their hands. However, those are indeed fillers as elucidated in an antecedent part of a series of events and outcomes between two major events of anyone's life.

There are a few people who free themselves from the "so-called fillers" by retaining intactness of true goal, which is significant and plays a vital part of life journey.

There is quite a paradoxical case of the complement of those few people, who are in a trap of their own fillers and getting out from such a trap invariably takes a longer time. Thus, the cyclic phenomenon of getting in and out continues perpetually until a significant event comes in. As we certainly belong to the latter category, if not all, at least a major portion, the world is filled with more complexity and the connected relationship of direct, indirect and artefacts supremacy over the natural existence.

The importance of fillers cannot be neglected and is non-complacent because of three fundamental conjectures. The foremost one is the True Goal of Life, which is a necessary filler with significance. Second, a very natural phenomenon of comparison and supremacy to others occurs between equivalent creatures ranging from mammals to reptilians or any other class. This comparison leads to various extended activities of a basic one. Thus, an interconnected activity called fillers come into one's life. The third is the necessity for the survival of the self and connected people, for example, a father who struggles for himself and his connected family member or a leader struggling for his community and followers. There is an absolute sense of necessity for such struggle or specific activities to be undertaken. Although those are fillers, there is a need for a filler with greater significance for humankind for the uplifting of future generations and guidance for expansion of respective genesis by means of selfless dedications. People live life with an intense and profound purpose instead of travelling in the Basket of Space and Time. The Basket of Space and Time is significant and possibly can be tracked for interpreting the direction of life, but How! To explicate such an idea, The God Point is introduced. So, let us start with the Universe subtly.

(Contd.)

The Universe

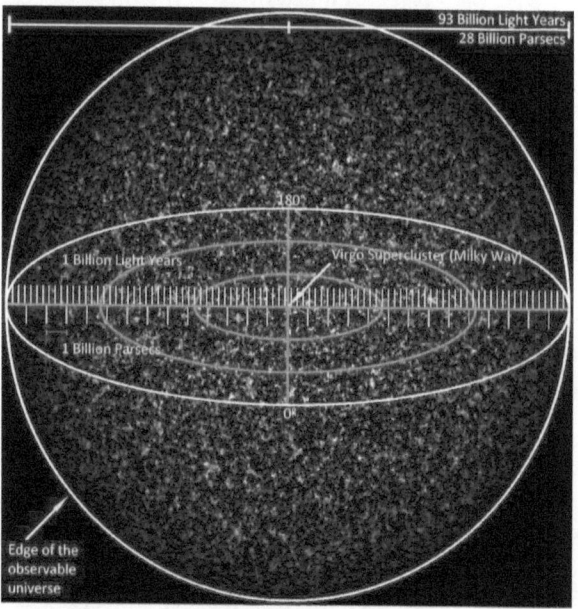

Can we see the Universe in one go or can anyone see infinity? This is an important and mysterious question, which can certainly be answered but with a small trick. Therefore, I request you to close your eyes before you instigate reading. The intactness of eyelid closure should last not more than a sixty seconds. When we close our eyes, we can see miracles and extraordinary ordinance. There is absolute darkness with infinity everywhere. The darkness prevails everywhere and indeed there are no borders. Yes, we can experience an infinite and darker state of the Pre-Universe Era of cosmology.

This is a pure state of emptiness, pure vacuum. The unconditional void is eventually filled with a lot of matter, which is the Universe today. In the vast Universe, I have tried to visualise from a point at the top of the Universe, The God Point. Let us assume that The God Point is located at (∞, ∞, ∞) which can witness all events of the Milky Way and Solar System in one go. The location is depicted below.

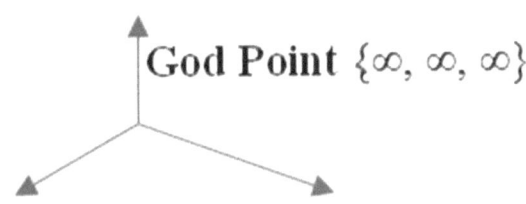

The point is imaginary and is taken as a reference for Incessant Surveillance and Reconnaissance for all pertaining notions evolving in chapters. The location of The God Point makes its presence ubiquitous with the ultra-gigapixel capability to cover almost all micro aspects as well. This is a purely fictional point without any scientific sanctity and is applied for cogitating at an ultra-dimensional perspective and unbiased earthly aspects. Thus, it makes sense when readers need to interpret instead of the physical identity of The God Point. It is obvious to be a possible concept, but it may need superlative intelligence and adaptive technologies, which may be encountered over a century or even more or may never be. With this note, let us move to a few more concepts of concentric cogitating and quasi-ideas.

When we talk of genesis and evolution, we are stuck with an important question, "Is evolution a filler or an important end?" Very innately, it is a natural filler belonging to the third category of fundamental conjectures. For any foremost society, which is to be unfolded with exploration to new adventures and procreation of potential outcomes, a leadership role and manifestation, however, cannot be neglected. Here is a catch between a leader and an influential and potential leader who is worthy of many voluntary followers.

The cynical nature of followers makes them involuntary and compulsive of being repellent, which introduces an opposition during the same time while a new and foremost society is evolving and unfolding as mentioned above. The same society thus forms an occurrence of Sun and rain concurrently. Hence, the second category of fillers was cynically introduced during genesis and civilisation. It depicts an illustrative concept of linkages between all three categories of filler fundamentals, thus unifying all fillers into one in one's life between two significant ends.

The fillers of life are natural, evolving, reliant on surrounding and obviously inevitable. This does not make a Human a great binding unless he/she is indulged in and trapped by pseudo-artefacts.

The stressful life thus deserves enormous fillers and twists instead of linear life without many intricacies. The urge to do, the chemical imbalance of the body and mental state are driving the next event of filler albeit being unpredictable because of surrounding involvements. Everything remains contingent on the thought process of individuals under different scenarios. For example, the way of thinking during a specific investigation enables one to suspect usually everyone who has closer inclosure with main events and their breakdowns. However, this will come to a very decisive state eventually after a cyclic thought process, basically numerous rounds of investigation followed by analysis and re-interrogation.

If we draw a graph of cyclic thought process with time, there is process recurrence thereby enacting redundancy in efforts but that seems mandatory because of filtration and fine-tuning of evidences. In such cases, to save time, one can follow the linear thought process; such approaches in life are good for limited instances but not every case owing to the imbrication of false evidence in the innermost layer of the overall investigation process, which is prompted by Humans themselves in view of protection and selfishness.

(Contd.)

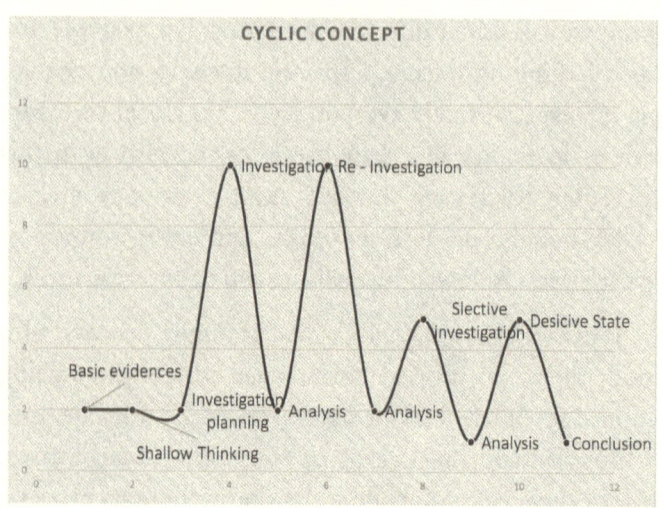

Linear concept

Both the processes are Fillers of Life but have different levels of complexity in the handling of the same case and the same conclusion. This depicts that Human behaviour is the cause of all complexity and stress in life.

Upon understanding such a concept illustrated by depiction, it can be devised that behaviour and subsequent thought process affect all the contemplations in Humans and events. There is a synchronic existence of another intellectual thought process to handle the fillers of life. This may be called the concept of concentric thoughts. Before stepping into concentric ideas, let us analyse one more thought process concept—the Hybrid Thinking Model.

The Hybrid Model is a combination of Cyclic and Linear concepts over a long-time frame. Let us combine both the graphs for a longer span of time, like a stint of current average life expectancy of Human for unbiased insights.

Looking at the mixed graph, there are sporadic phases, which when combined and compared over long stint instead of short to medium span look something depicted below. This exemplifies that the Hybrid graph spawning over a long span proximally follows (usually) a linear trend. Thus, for the overall life span of Humans, a linear trend is followed. All the differences and singularities followed by indiscretions and intricacies are subsided in general and abysmally visible subject to magnification with respect to *shorter intervals*. Once we extend our thoughts beyond *those intervals* aka *Fillers of Life,* everyone's maturity level emulates at a certain point of time. The state of Space and Time comprising alignment and quasi-thinking ability among people merges individuals to groups and transforms complexities to Linearity. After all, we are *Homo sapiens*—one group and unique breed.

The takeaway from such a concept is to emphasise that Fillers of Life are taxing, complex and typically sustain for shorter intervals of life span. Those are not eternal and should not be signified as crucial as a state of Life and Death or any significant events. People often say, "take it easy," and yeah, they are right. When events are created by imbrications, it is very natural to be affected intensely by a past instead of the present scenario. Such events aggregate to be Fillers of Life, which have no implications unless they sometimes abysmally lead to some significant events, which may be undesirable or desirable. The Hybrid Thinking model throws light on Human thought analysis over a different span or ranges of time. This is nothing but a play of time and events. Therefore, "The Play" must cross us but will never make us strife profoundly if we are determined, courageous and judicious enough to be called mature personalities. Let us move to the idea of concentric thinking.

(Contd.)

The Rewind Mode

"The rewinding of Universe (Journey to Big Bang)"

Through

"Pulling in concept"

The Universe is full of Entropy and Randomness. The occurrences of events are certain; however, things are intricate than being customary and sometimes beyond our imagination.

Entropy: This is the measure of randomness of the system. To make it more comprehensive, let us understand Entropy in very simple terms. "The Entropy is *disturbances of a system*" and that is the fundamental reason of Entropy translation into eventual uncertainty of occurrence in a diverging manner.

The pulling in concept is depicted in the figure, where all the planets are pulled back to a single axis. This will make us better comprehend how singularity is contributed in the formation of the Universe through Big Bang. The pulling in concept is about reiteration of the idea of Big Bang and going into our chapters through The God Point, which focuses primarily on four major components viz Symmetry, Balance, Energy and Entropy.

The God Point views the Universe as depicted in the figure. This book is largely based on the Universe and life formation concepts/idea through the imaginary point located at infinite coordinates "The God Point".

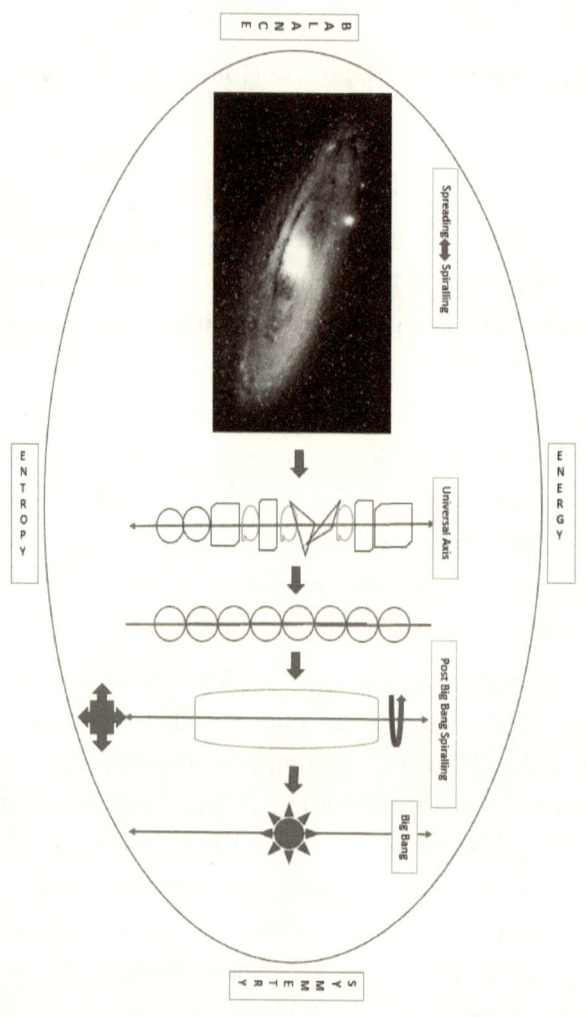

Retrospection of Big Bang – The Rewind Mode – The Rewind Concept to emphasize existence Intellection is indeed four Major aspects; Energy, Entropy, Symmetry & Balance and their Transitions move this Universe. (Please Turn 90 Degrees) "Anticlockwise to see retrospection illustration"

Energy, Entropy, Symmetry and Balance are Integrity of a Universe

Upcoming chapters explicitly cover all pivotal aspects of Life and Universe.

- Fillers and Entropy
- Role of Symmetry
- Balance
- Energy Transition

The pattern of the Universe is random and sturdy spiral and it never ceases expansion. While this remains the Universal truth, there is an unceasing retrospection leading to mystified and disguised reality, which Humans can never ascertain with accuracy. Quantum studies and analysis revealed some persuading aspects about a random beginning like Big Bang and similar principles. Considering those, I have a concept of merging and demerging of Symmetry and Balance, which can be a probable cause of the beginning and expansion of the Universe.

To translate this idea, let us start the rewinding of Time and Space from now to a very nascent state. This requires a certain amount of fantasy envisaging by readers. Assume all the masses of the Universe, which comprise planets, stars, stones and any solid matter, to be in one axis, one over another. This creates a structural Symmetry, apparently like a chain of beads posing singular axis and vertical masses with zero slant. Let us term this process as a reunion of celestial bodies.

For illustration, the stagewise depiction is presented here.

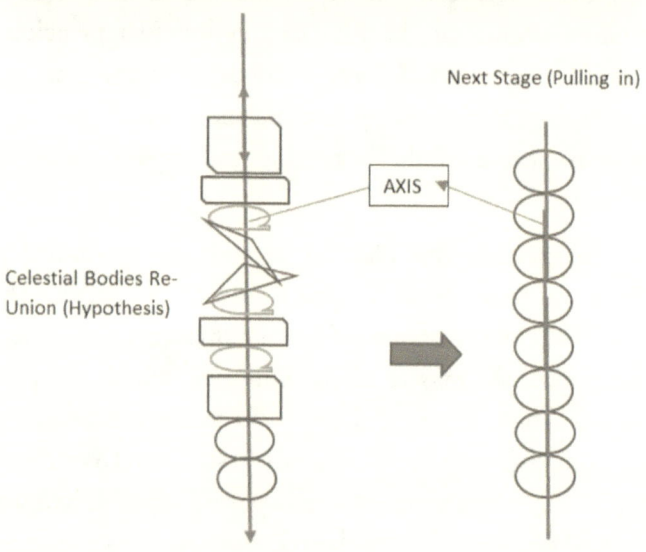

The re-union of celestial bodies when travel against Space and Time, they all will be merged into a Dot. The Dot is the "Point of Big Bang"—when the big bang started. This concept is called Retrospection of Big Bang put forward for understanding Zero Point or The Primary God Point. Eventually the Galaxies and the Solar System expanded from Zero Point and now we are so far from the Point. So, keeping the Earth at Zero, The God Point is at Infinite coordinates of Universe/s.

Takeaways from The Rewind Mode Sub-Chapter

✦ Four Important Enablers for the Universe :-
   1) Energy
   2) Entropy
   3) Symmetry of Matter inside Universe
   4) Process of Balancing
✦ Terms to remember
   1) The God Point (Zero or Infinite - Relatively)
   2) The Retrospection of Big Bang
   3) **"Pulling in"** – Re Union of Celestial Bodies
   4) The Merging Dot or The Big Bang Point

Let's start the next Chapter with the Concept of Symmetry…

ROLE OF SYMMETRY

It was often felt by many people living on Earth that the entire creatures in the world are created by some special entity that has tremendous energy and adequate powers to do miracles. The most common and important point to be noted is that "all creatures share a similar property of similarities in their formation and stages of structural development", which certainly generates a non-negligible amount of curiosity and belief on such presumptions in one's life.

Such things never existed or have never been proved scientifically, but a billion experiences were shared by a million people in the above context, which polarise people to think recurrently and coherently. Hence, presumptions will continue to unfold further because of persistent thoughts and discussions, sharing of experiences and pertaining experiments if any. However, the conclusion is always setting itself far from assumptions and convoluting ideas.

As a reader, can you guess the similarities among every creature on this Earth?

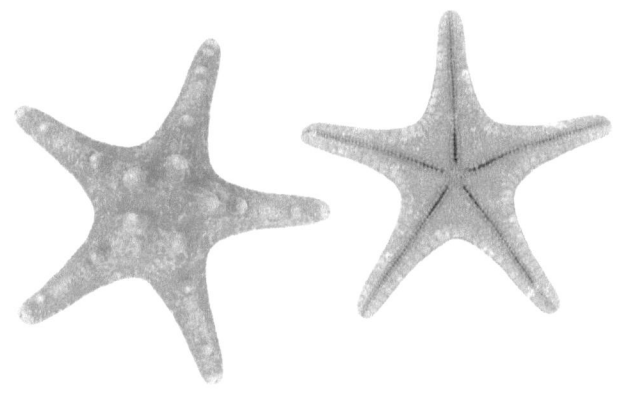

Symmetry

Yes, that is the absolute answer if you have the right guess. Let it be radial or bilateral under existence. It does not mean that Symmetry existed everywhere without exception like asymmetrical unicellular organisms. However, multicellular and eukaryotic organisms are the perspectives of Symmetry for the ongoing chapter.

The symmetrical shapes and their existences always made Humans for being compulsive to think on creation by a Supreme entity. A question always rings through my mind, *How can such perfections occur naturally and involuntarily?* To analyse such a concept, one should identify roots and get into explicitness of anatomy of any organism.

The anatomy of any mammal is to be studied explicitly in the purview of its basic formations. Such study will flash about similarities and dissimilarities, which arise as a part of the evolution process. Then comes the role of the cell, the basic unit of life, in the formation of life on Earth and extrapolation of the micro kingdom.

How do fillers pertain to Symmetry? Any filler is stated as a random activity performed in a fashion and uniqueness. However, when the micro-level similarity is neglected for a while, we can see that fillers follow a definite and repetitive pattern, which is merely related to object Symmetry. For example, when we undertake the responsibility of being parents in our life, right from reproduction to birth of offspring and then parental care, especially a mother's care, which is the absoluteness one receives, it may be once in a lifetime because of the uniqueness of a specific mother and her child. This is common among all mammalian mothers. The care follows a similar pattern of what a specific mother has received during her childhood or else in tune with thoughts of her missing hood, if she is an orphan unfortunately or not brought up with a desired level of affection. The upbringing of any child will follow a specific pattern, from days to months, months to years and so on… until adulthood is attained. Such a pattern is closely related to the Symmetry of Human Beings; we have forelimbs, so the growth pattern differs and that is why the caring pattern will be naturally altered and revisable.

> We behave analogous to Symmetry in our anatomy. The foremost behaviour, which is closely related to our Symmetry, is BALANCE. The Balance factor is embodied in every activity, even though it is a random one.
>
> We always try to keep Balance, whether being in a conscious or subconscious state of mind. Now let us evaluate some more examples in terms of Balance and Symmetry leading to alterations to a filler, which was supposed to be carried out in a different pattern. Take an example of a milestone of playing and winning a football game. This game has both Symmetry and Balance, as the ground is symmetrical in shape and players have Balance and activity of securing goals, needless to mention that any imbalance in the game will lead to success or failure. If we tend to Balance the activities, the game will continue until there is a significant end. This is how life is… "numerous mini battles of Balancing and finally ending up someday".

This is how Symmetry affects our behaviour and micro- to macro-level activities. For example, an election of any country is, of course, a game of balancing the seats of its parliament. Our anatomical Symmetry makes our brain so symmetrical in shape and its performance that it always tries to keep balancing the activities we do following a desired pattern. Here at this point, the difference between randomness and Symmetry and Balance is further explicated.

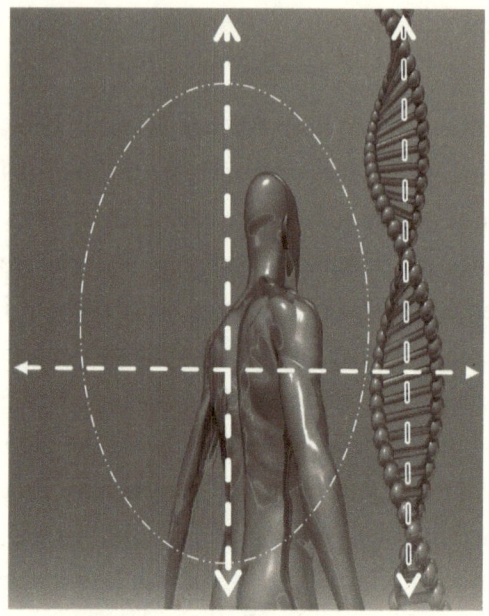

THE HUMAN SYMMETRY

This is because randomness exists everywhere, but it cannot be without certain goal of balancing. Sometimes, truly randomness has no goal and moves endlessly without repeating any pattern, but that has an end. Such an ending may not be balancing but the entire process will be carried out in a very balanced manner involuntarily and without any cognizance of micro-pattern of individual activities.

Fillers in life are repetitive most of the time and follow a pattern. Ironically, *Homo sapiens* follow them despite irrational redundancy sometimes. If someone ever tried to avoid redundancy and does every step with a unique approach, then consequences are either a big failure or a lukewarm success. This is because organisms are designed to follow repetitive activities inevitably throughout their life span; hence, repetitiveness is the basic necessity of survival. There are few processes, which deserve to be correct examples like respiration, photosynthesis, digestion, cell divisions, etc. This is even common with non-living entities, in addition to the living entities (plants and animals) on Earth. The rotation of Earth, sunrise, sunset, water cycle and weather change pattern all are repetitive processes, which are ultimate showcase of a Pattern, Symmetry and true Balance.

The cyclic nature of fillers is inevitable as far as relevant processes are considered. However, this is not the conclusion, just an intermediate derivation of a brief study on pattern and repetitiveness of entities with respect to confined processes. The terminology "intermediate derivation" in the above para is contradictory to Symmetry and cyclic nature, which indicates awaking of new testimonial on ceasing repetitiveness. Yes, *Homo sapiens* should stop redundancy of many activities in life, thereby compacting the fillers in life and eliminating fluff thereof. Such elimination is an uphill task if the situation is dealt with incoherently and inappropriately. This is because restoration takes a lot of time and sometimes can never be done if eliminated without any preparedness.

The subsequent chapters will provide guidance on compaction of fillers in life. Therefore, at this point, this elaboration can be terminated and restarted in the next chapters.

Symmetry introduced Balance, Cyclic Nature and Pattern of Human Behaviour. The next chapter will cover Anti-Gravity Effect—Balance and its role on Fillers of Life.

Life is about something in which everything that is encapsulated with boundaries, which always tries to afloat between borders, plays a role with and without physical interaction where unknown affects more than known. Therefore, knowing is not that crucial than pragmatic predictions and capability building to face uncertainties to a greater extent. The symmetrical aspects of the Universe, of course, a hypothesis, cannot be ruled out because infinite creation cannot be witnessed unless something faster than light takes over Human biological age to encounter multiple galaxies and Universe enclosure point. One thing is for certain that the Universe cannot be measured; this is because the Universe is an infinity. What we call a Universe is, in fact, a complementary Universe and the real Universe is Supreme and superset of its complement where we reside at a negligible level of existence with respect to size. The larger the deal, the greater the efforts to put on, else things will be in the soup forever. The Universe is in a matter of fact of infinite dark matter, which eventually encounters action and reaction for a portion and not as a whole. This contradicts the virtue of posing a beginning and end because no beginning can be ascertained for pre-existence. The part or the complement of the Universe is the current Universe of existence, so it is left for the contemplation of readers to call the Universe or its part. If galaxies are not the Universe, then the whole Universe today never equates to the real Universe in any aspect. The parameters Randomness, Symmetry and Balance must have instigated the beginning of fusion, fission, dispersion and an uncontrolled and Entropy-filled spread of the Universe. The Big Bang is not an instant event, but resultant of multiple parameters as mentioned above that existed and acted concurrently. The conjecture of God factors should start from here and not from the Solar System, which is often perplexed and misinterpreted by grand great ancestors.

(Contd.)

The God Point

Retrospection of Big Bang – The Rewind Mode – The Rewind Concept to emphasize existence Intellection is indeed four Major aspects; Energy, Entropy, Symmetry & Balance and their Transitions move this Universe. (Please Turn 90 Degrees)

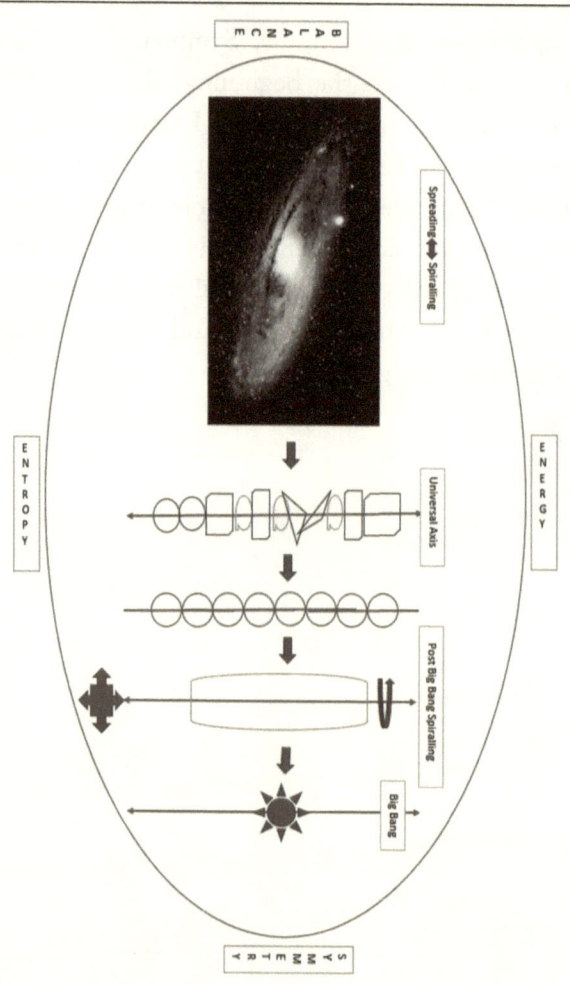

THE ANTI-GRAVITY EFFECT

—⋅⋅⋅✯⇤❁⇥✯⋅⋅⋅—

The balancing process is the basis for the formation of Earth and life. The BALANCE started from the Universe, right from Force. The Force, which can be either a pull or a push, is spreading throughout the Universe since the inception day of Big Bang. Once the force applies, there is no further way to stop dissipation of energy. However, better channelisation can tap the untapped potential within it.

The balancing indeed is a phenomenon generated since the inception of Big Bang. However, what caused the balancing act like Big Bang in the Universe is still unknown in a scientific manner; however, lots of hypotheses have been proposed. If something has ever existed before the Big Bang with a certain amount of force and energy, is that an ending of the previous Universe or an era of cosmology? Well, that is a cyclic question with a never proven answer. A stone somehow collected and went on carbon dating and confirmed to be dated before the Big Bang will lead us to hypothetical, inexplicable and intricate thoughts of the evolution of Force and Balance obviously. Nevertheless, we are fortunate enough to have a life formation on Earth by whatever means. I have briefly stated the role of balancing since the inception of the Universe; further elaboration will follow in the current chapter.

Technically, the Anti-Gravity Effect is a countervailing virtue to Earth's gravitational pull; this may be a satellite launching or staying in a vacuum. The Anti-Gravity within Earth's atmosphere induces a loss of balance to physical forms, thereby disrupting symmetrical establishments and processes in organisms. It is to be self-elucidated that organisms cannot survive in Natural and Unplanned Anti-Gravity mode for much of their time owing to counter mechanisms and strive against the radical design of their composite structure of Balance and Symmetry. That may be a probable cause of the existence of life on Earth—the Gravitation. However, an astronaut's case is seemingly looking to be normal while being in an Anti-Gravity mode because it is pre-arranged by artefacts and somehow not going to modify an organism's established characteristics. This is purely a Human attempt while not considering any symmetrical formations generated from space. If at all any experiment would ever be conducted from an International Space Station by developing third-generation animal's zygote to fully formed baby and the gradual growth observation in the Anti-Gravity mode, that should ideally give accurate and pragmatic information about the role of gravity. The third-generation experiment here is defined as follows: animals should be scientifically life supported and allowed to stay until leading their offspring chain up to third-generation species, which were born in space and thus the experiment should start to watch the growth pattern under Anti-Gravity conditions with all enabling life support and food arrangements. If required, any scientific tools or arrangements can be used to make vital life processes functioning in an artificially natural way, but seemingly as absolutely natural. For Humans at this juncture is a jittery case for such unethical and a time-enervating experiment. However, it should be clarified profoundly that the Human structure should not have greater change due to pre-coded growth pattern inside genes. However, this is not the case of an early organism in which genes do not carry pre-defined memory.

Thus, they are bound to be affected by the role of gravity and certainly structural deformations or asymmetry or an undefined shape to be established. With a closure note to the above topic of Symmetry, Balance and Gravitation, a brief and envisaged conclusion is mentioned in boldface for reader's viewpoint and extrapolation of aforementioned hypothetical, theoretical and undemonstrated concept in general sense.

Ideally speaking, Symmetry is formed because of Balance. The Balance is because of gravity of Earth and any kind of pull for other planets. Symmetry not only exists on Earth; it exists in the Solar System and Milky Way when viewed at a supreme level of watchpoint of a Universe.

The God Point can further unfold a few mysteries of a black hole and invisible bodies in the Universe.

The galaxies, black holes contained inside a Universe, give us an idea of a Universe. However, if we extend our thinking beyond a point of the Universe, a hypothetical model of **Multiple Universes** will come under consideration.

This is where the existence of Multiple Universes as a concept strikes into mind, which may cast a serious doubt and is open for a scientific inquiry and gamut of research. Life is all about exploration; it follows either a systematic or an unsystematic approach, which all depend on the necessity and the thought process of any community. The usage of exploration in the above line is imposed for extending thoughts on Balance and Symmetry, which are articulating this chapter with Life fillers. The book is about a briefing on fillers and The God Point where understanding is unbiased and unanimous alongside the coherence deduction of facts with respect to event chronology. Thinking profoundly on the subject matter will make readers feel nigher with The God Point and synchronisation of their thoughts.

The balancing in an organism is due to Gravitational Pull and Solar System as a whole. In view of balancing, organisms need to attain Symmetry. This is where Symmetry comes into the picture in a broader sense; however, a deep dive into the insight of anatomy and function will further elaborate on its self-perpetuating balancing process. Technically, Humans possess symmetrical structure from an outer view, i.e., a view of the surrounding; however, Human structure is asymmetrical from the inner portion fundamentally in view of their endodermal formations such as internal organs; why is this so? To understand this, let us enlist the radical requirements of Human Beings on this Earth. As per religious belief, Humans need five basic elements as depicted below.

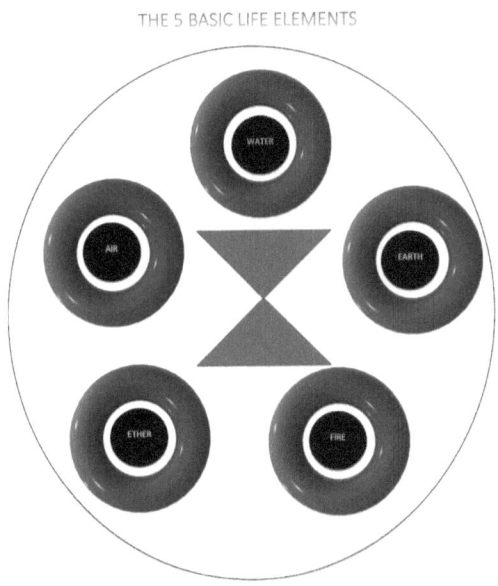

THE 5 BASIC LIFE ELEMENTS

To make it more technically acceptable, let us write them as Air, Light, Water, Food, and Soil which is fertile. If any of these five particulars vanish because of any reason, the survival probability of Humans diminishes gradually or instantly depends on in which order they start to deplete. (Air for Oxygen, Light for Photosynthesis, Soil for plantations and dwelling support with the ecosystem.) The balancing is essentially accentuated internally and externally. The internal balancing, however, cannot be visible but can be demonstrated with suitable examples, fluid balance in a body, cell divisions, heart balancing two kinds of blood within the body, lung's gas balance and many more. Thus, balancing and Symmetry occur concurrently in majority cases, let that be internal or external.

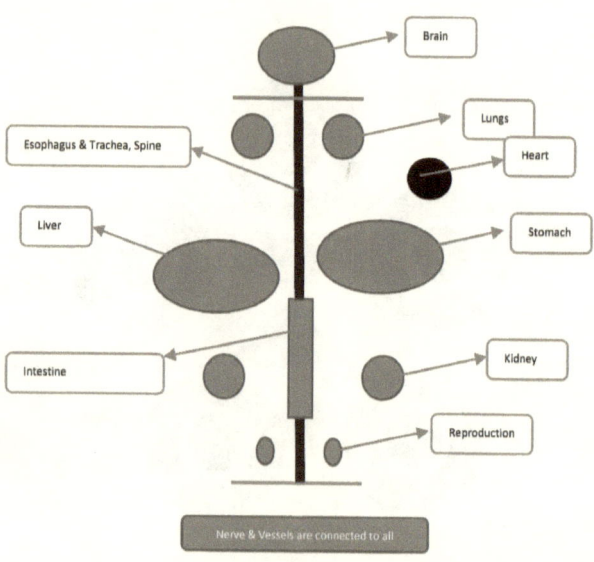

The depicted illustration is about the Human body, which often needs Balance in every aspect. Upon deducing further facts from the illustration, we can radically see a long tube right from top to bottom. Eventually, we might have posed bilateral organ development as per the requirement. Thus, Symmetry and Balance are there since the commencement and they continue to exist.

Although the evolution is a widespread generic interpretation, it has never been completely understood and the gamut of research is still going on owing to Mysterious eventualities and discoveries in every stage of Human development in the 21st Century. Thus, I have an extended idea in evolution. The important purpose of evolution has a greater purpose, which has been mentioned in the subsection of chapters. However, few paces are depicted as a foundation of understanding in successive chapters and insights.

Imagination and Knowledge

What we know is termed knowledge, and our thinking ability is termed imagination. Everybody has knowledge on his or her respective area as per the exposure and ability. Nobody has entire knowledge in a subject because of infinite content, which needs an imaginative approach to decipher the long-standing mysteries. "KNOWLEDGE is like winding, which has to unwind with a tool called IMAGINATION on everything."

Role of Balancing and Symmetry in Fillers of Life

As stated in the previous chapter, "It was often felt by many people living on Earth that the entire creatures in the world are definitely created by some special entity that has tremendous energy and adequate powers to do miracles." The dilemma created by Symmetry in the structural formation may be true or just desirable thoughts, which can never be wiped off from the Human mind because of the upbringing pattern and religious indictment.

The fillers of life start from basic survival and slowly extend to the growth phase.

The role of Symmetry and Balance is everywhere in life, right from zygote to various stages of one's life. To understand the significance of life events and in-depth analysis of life, one must watch himself/herself from The God Point. The God Point, which is independent of gravity and other earthy elements, clearly specifies what is important in life, how that is important and what is the responsibility to be undertaken in life by *Homo sapiens*. Let us assume that someone miraculously reached The God Point and started demystifying various activities and causes. The explicit description will follow by a case study, which is again hypothetical and imaginary for knowing life and its fillers in absolute form instead of relative factors under irrational analogy and biased approach. This will further altercate formulaic thoughts and unfold the true sense of reality and absoluteness. Every creature on this Earth bestowed with a specified living pattern, a natural support, is substantial on this matter. Before stepping into a case study, the life expressions of living entities are to be understood. Organisms often exhibit similar virtues like Reproduction, Protectiveness for self and offsprings, Curiosity, Anger and Emotions (Subjective). Although the presence of such qualities may be unique for specified organisms, in common, they all follow a pattern of life expression.

Let us decode the life of *Homo sapiens* right from a vibrant, credible and forefront sperm entering into a capable egg inside a Uterus. If someone tracks the cellular formations stringently and incessantly, he/she will be amazed to see each clip depicting various transformations occurring sequentially as if a specified divisional algorithm was pre-coded. However, it is absolutely a very natural process without any pre-defined algorithm. The amusement does not last for much time and ends with curiosity followed by multiple perplexities with correlated factual deductions.

Here comes germinative curiosity on the existence of a pre-dated and pre-defined memory with biological sequencing in the sperm or egg or zygote.

What is the role of the Universe at the moment with above?

Anyway being a case study, someone residing at The God Point can see how exactly this occurs. Is there any wave or radiation-influencing zygote or baby-life generating innately and automatically with an internal control mechanism? Please think for a while.

A post-cellular division of the micro zygote, which enters into a critical phase for a week and subsequently the blastocyst, occurs. Then comes most key aspects of Human survival and vital organs like the heart and its formation. The heart beats, thereby supporting gradual formations of required organs in Human. Once embryo renamed as a baby is delivered to Earth, the actual life starts in true sense.

The Human brain develops slowly by keen observations to its surrounding. Development of the Human brain is the most crucial part of an organism owing to close interaction with parents, society and Earth. The developmental phase should not be lacuna of basic elements and at the same time no overdose induction in any field; this will ensure child's brain and resulted thoughts inclination not being slanted and being thoroughly neutral. The growth should be qualitative, natural and involuntary. However, nobody can make this possible because of pre-existing scenarios or circumstances and distractions in day-to-day life, lack of a dedicated approach filled with concentration in every move. Going back to the case study from this point will repoint the cursor to the developmental phase of *Homo sapiens*. This discussion is concerned with the one and only category of mammal, i.e., *Homo sapiens*, and should not be perplexed with evolutionary developmental phases, which will be discussed later in upcoming chapters.

The God Point Observations would further inscribe into insights of emotional developments subsequent to the brain development of a case. The emotional quotient is nevertheless most significant and idealistic during the nascent stage; however, that turns up slanted as the time proceeds with respect to incessant day-to-day interactions. No opinion is fixed or eternal except some time lag or lead between connections.

To vet and retort any single persuasion before it is extrapolated into generic ideas and level-headed analysis, an introduction of a difference of opinion, a second comparative case, must be inducted for observations. Hence, the observer (The God Point) will have two cases for parallel study; the Synchronic cases being observed concurrently and in a stealthy manner to locate minor changes at the bpm level (as stated in the first chapter, bpm level, billion frames per second).

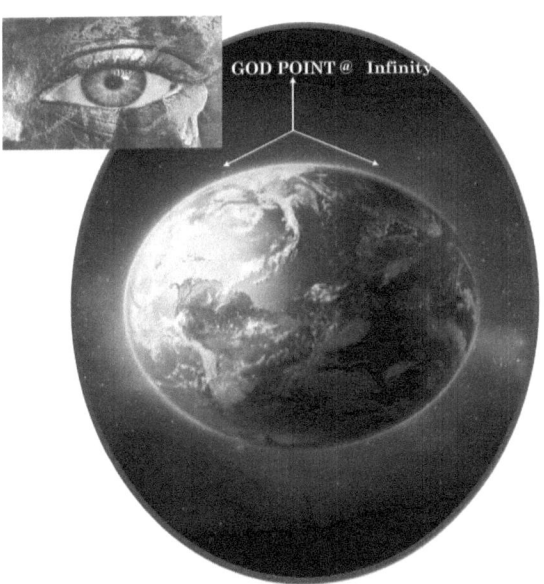

The growth patterns of cases, however, merely similar on the physiological level, differ owing to variations in their event chronology. This Universal truth strongly upholds its position even for initiation of cases at the indistinguishable point on Earth. The exemplary cases of twin's birth are not exceptional because of the same dimensional positioning and with a micro time difference. All the subsequent events are naturally linked to their predecessor and those subsequent events will further lead to another set of events.

The God Point entity can finitely distinguish the two cases with respect to their behaviour, responsiveness, handling stress, joy, emotional control and activities as well. However, this entity cannot know what is going inside their mind, which is, of course, a fact. Like a live camera within an orbiting satellite, The God Point will observe everything, but the factual informations are independent of the Earth and the Milky Way in any manner. Surprisingly the behaviour patterns of the two cases look similar when both are exactly opposite in terms of exhibiting actions. For example, let us consider the behaviour of two teams in a game or competition or war. The ultima is to win as a team by exhibiting counteractions to the opponent. The opponent's reaction is similar when it was looked at the other side. Both exhibit similarity for countering activities. Thus, the balancing factor comes in recurrently, as this seems to be an organised event. In the case of unorganised events, the balancing may or may not be witnessed. To understand the insight view, let us take another example. Two cases of a surgeon performing surgery and the patient who is undergoing surgery. To be very specific to case lets dissect this instantaneous and few previous events into 30 important frames. There is a similarity of worries, beliefs, confidence and care to cure being captured in frames. Even though two persons are pole apart and unfamiliar with each other, they are very synchronous. Although the surgery was started without any intention or clear plan of surgery, i.e., in an unorganised fashion, the ending was as if it was organised and decided beforehand. Here comes any unorganised task routing to the balancing approach by reorganising itself. That is the beauty of nature. For being double certitude and thought refinement, let us add up one more citation; consider a case of an accidental meeting between two people and subsequently, such meeting transcends into a longer relationship. To understand this better, we need to check clips by reversal of event occurrence at a certain point of time from The God Point; once reversal scenes are observed, nowhere The God Point can estimate these people are meeting at a particular point of time. When reversal-based observations of such an event were taken back to two decades, the occurrence of a current event would absolutely be unpredictable.

Let us consider this point as the most unorganised and unpredicted Life Fillers of the people who are going to encounter after two decades. The "organised" here means the aforementioned example of two people facing each other for the first time, like never before. How does an unorganised event turn into an organised event and what is the role of Balance there?

To answer the above question, we must understand event chronology. What could be the first step of the most unorganised random event? (i.e., when both the cases were born). Let all these events be observed closely from The God Point. The primitive steps are basic care and upbringing pattern of children by parents. Subsequent events would be growth stages and exposure to surrounding factors, where lifestyle, society and crazy things suffuse into brain involuntarily. These random developments make individuals think of balancing—the balancing of emotions. Good friends and bad friends, selfish and thanksgiving days, helping pattern, showing care and anger, expressing desire and passion alongside frustration, irritations and attractions—finally a Human is built with enablers of balancing of the self and surrounding.

Progressively Humans are curious about their preferred partner and friends. Once the preference is established, the directions will flow in a desirable way despite a certain amount of resistance from randomness. The randomness represents unorganised event sequences and the atmospheric, geological, and environmental variations, which affect the lifestyle and behaviour of humankind. For example, the summer days and rainy days will influence health and event sequence. Ironically, all randomness comprises earthy elements, which must balance themselves in any means to restore our planet and save the liveability on Earth for all living organisms. The influencing factor here is society and internal factors and randomness refer to uncontrollable and unplanned events occurring in day-to-day life.

Finally, an organised day will certainly occur, if failed, and then re-occur, but the occurrence is certain.

Therefore, the role of gravitation is important in Symmetry and Balance of any events. The gravitation induces Symmetry and embeds balanced thoughts in genetics and organism evolution thereof. Let us think beyond Human and extend the analysis by studying plants, other aquatic and terrestrial plant or animal bodies. The growth pattern of all the trees is similar because of the Photo Energy requirement and their survival. Such aspects are not only applicable to plants but also animals as well. For example, the aquatic animals do not possess feet due to the absence of terrestrial balance of gravitation and requirement of moving in water through tails.

(Sometimes, magically in a few aspects, where intentionally some events were tried to change by means of time travel, people will get close again due to inherited behaviour and emulation. This is again hypothetical and is not applicable because of their rarest occurrence.)

Indeed, this is a widespread fact of evolution proposed by a lot of scientists, from Darwin to many scientists, postulators and scholars and any pertaining group of people. Let us generalise all creatures into multicellular organisms and their diversification pattern. Post generalisation, we can observe a reactive nature of the organism to external factors, which affects an organism's behavioural pattern. Such behavioural pattern may be present for a shorter term or longer term, depending on the exhibition pattern of the respective external factors. The external factors, which influence the behaviour of multicellular organisms, may be steady, jolty or fluctuating. The Earth, which is a major portion of external factors, has kept itself in the best possible balancing way by enabling sustainability of life on it and living things thereof. Undoubtedly, we are fortunate enough for the ongoing rotational stability of the Earth, Sun and Solar System. The responsiveness of an organism may or may not reciprocate in a synchronic manner. The primary cause of unpredictable responsiveness is the experience of an organism. The experience and responsiveness are closely interrelated for any creatures in the world. Thus, any experience in life is significant as it infuses some memory and decision on the response pattern. Sometimes the external factor based on experience seems vague but that must not be neglected or overemphasised.

Interestingly, responsiveness may be involuntary, voluntary, or voluntarily involuntary. The fillers of life have an infinite number of encounters to experiences and responses.

Any behaviour pattern is an overall result of experience and responsiveness. The Homo sapiens exhibit a typical and optimised behavioural pattern, which has a greater significance in Life Fillers as they drive the next event in a voluntary or voluntarily involuntary manner. The behavioural pattern is totally optimised over a period of time; for example, children's responsiveness and experience make them optimise their behaviour to the same event in a different manner as compared to adults; in general, the difference can be further widened with adult and older population. Such example can be further extended to the behavioural similarity among the same class of animals and dissimilarities as well, for example, a pet dog and a street dog responding to the same event.

The gravitation ensures the stability of life by stabilising water bodies, atmosphere, organism balancing pattern and protecting from other planer intervention within its cosmic territory. Life Origination, although took place million years ago, has occurred because of massive benefits from the Universe and Earth generously. The gravitation G = Gravitational constant × masses again varies with respect to the planet. As gravitation is not a sole factor for life generation, other planets need other life-enabling factors to be available and act for the generation of life. This completes the chapters of Gravitation and Anti-Gravity Effect.

Thoughts to come forward:

+ The concept of an artificial planet comes as a new idea for the survival of life alongside the Earth orbit other than searching for Mars, which is still inhabitable and a reconnoitring option.
+ Organism Structure without gravity and
+ Organisms existence without carbon and organics.

Readers may require thinking on these concepts in the purview of role quantification of "supernatural power" to explain the existence of life on Earth; however, this is a suggestion, not an affirmation. In such a random world, any event happens at any time, which is certain to be unplanned most of the time as per the recorded observations from The God Point.

Summary: The Gravitation introduced Balance and Symmetry. Symmetry might be the most vital factor in the formation of life. The transformation is based on functional and survival requirement on prima facie; the uniqueness of every organism ensures overall diversity and Balance of the Green planet or Bio planet like Earth. Although the majority of space planet is still under exploration, so far Earth is the only fit planet to pose required virtues of liveability and safety.

On a lighter note, even the Earth is not suitable for life because of geographical and climatic aspects; however, the possibility of a cent per cent liveability planet for the major surface area can never be ruled out. Either the planet or the organism will adapt to enhance survival and biodiversity for the currently unfit and unexplored areas. Such an act may take 10000 years to one million years from now. On the contrary, the climatic conditions can never be in any allowable range like the current era; thus, the possibility of unceasingly supporting the extension of life on Earth by current Life Enablers of the planet is abruptly or progressively ended by the aforesaid range of 10000 to one million years. This all depends on many ongoing actions and reactions in the Universe, especially the Solar System. Fusion and Fission may be the most significant processes among all energy spread, but ultimately it is the Energy that runs life.

MYSTERIOUS LIFE

"Life is beyond time, flesh and bones, the imaginative capacity of Human brain."

Life, which has instigated from iterative organic manifestation, is an instant spark within an organism and its origin is the stupendous beginning of lively Earth. The origin of life owes to random and iterative interactions of organic elements within optimised conditions of pressure and temperature of the atmosphere. Thus, there was no magic bullet trigger for the generation of life due to iterative and random nature of unlimited possible combinations spread over infinite time on earth. The interactions were among organic and organic elements and organic and inorganic elements within controlled natural factors. The enablers of the life formation process inadvertently supported the process. The Solar System plays a vital role by the incessant supply of Solar Light with optimised parameters like temperature and intensity. The Solar Gravitation pull has been keeping our planet's rotation intact ensuring collision-free planets of concentric orbits. This is the pattern of the Solar System of the Milky Way; however, other galaxies may or may not pose similarities, which is still to be reconnoitred.

The timeframe of life formation indeed was very long, unrestricted and unbeatable without any defined milestone. Although the formation process had consumed an enormous amount of time and energy, it bestowed *Homo sapiens* with the utmost, invaluable and phenomenal creation of the Universe, the life. My thoughts from The God Point are that the chronological description of all events occurred in epochs right from the beginning to ongoing. Previous chapters were about the Symmetry, Balance and Gravity. In this chapter, let us discuss on before life and post-life aspects. Such aspects are a matter of Human sensitivity. Hence I declare this is solely my conjecture, not concrete or proven information. (This is neither religious nor scientific, just an intuition based on pragmatic conditions.)

A mystery always lies a mystery until its status is blown away by factual information with evidence. Even though a plethora of evidence has been available, they have never been adequate for deducing the facts and proofs. Hence, any concepts ever presented to clarify, justify, or criticise are baseless and that is why life is always mysterious before and the aftermath. Thus, as an attempt to unravel a mystery, we can never ensure proof adequacy, but theories can be postulated, or a brief idea can be proposed.

In line with this, I put my ideas in successive paragraphs. However, those are not for agreement or disagreement but for aligning and extrapolating the readers' thought process in general sense. Although there are no upholding mysteries, the realities will supersede superstition and other contrary beliefs, thereby enhancing qualitative utilisation of time of the new world. Thus, the entire world will be into forward-thinking instead of retrospection.

The religious rituals and beliefs of our ancestors and their worship dedication always helped us rethink wider prospects and draw an invisible and distinguishable finite line between two social opinion coverages. The people might have accrued through inconclusive discussions; a variety of people who meet during various stages of life will keep the thoughts steady, intact and refined instead of volatility. To progress the view on afterlife events, let us position our virtual self at The God Point to think beyond earthy and known prospects. In addition to virtual and imaginary positioning, we need a stronger penetration into before events. As quoted by Einstein, "Imagination is more important than knowledge." We are progressing in line with the quote for answering important questions of Life.

Let us uncover the before life part initially. The ultimate question comes "*__why create life?__*". The answer is "it's an unplanned event that happens eventually" with or without intention. However, there is a negligible role of anyone in organising an unorganised event; this is because of the spread of randomness as clarified in previous chapters. We cannot plan future even though our planning is always to execute the future. Let us get into insights of some intriguing questions and credible answers although they are conceptual.

The God Point

WHY
- *Why is life Created? If somehow life was created, does this creation deserve any purpose?*

WHAT
- *What happens after life?*
- *What happens before life?*

HOW
- *Is the Life Imbrications being altered?*
- *How the transition of life goes on eventually?*

GOD
- *How is the Supreme God's influence linked to it?*
- *Is it required to worry about afterlife?*
- *Is it really important to be spiritual and enlightened?*

WHY

The astonishing fact of the word "Why" is,

"It evokes a sense of curiosity, which embraces happiness or anger or displeasure or anomalous situation before its usage through an expression."

The subchapter "Why" is a conceptual approach of creation put into, which is solely based on my opinion and readers to follow in line with their thoughts and take away the parts or as a whole.

(Contd.)

Why is life created? If somehow life was created, does this creation deserve any purpose?

The responsiveness is to stifle while reverting with the persuasive reply to such a stringy query. However, numerous unverified answers were found in religious testimonials. While science has its own theories and contradictions, religious beliefs within people created generalised and deliberated concepts of the supreme entity. There is nothing wrong in having respective ideologies until deeds are being executed in errand of humanity and liveability on Earth. In an average span of sixty years, people can witness and respond with respect to their concerned dwelling era, which indeed restricted their thoughts to certain ideologies. This can be further extended towards definite reasoning from The God Point.

The God Point must have a credulous answer, at least theoretically, as it has kept its incessant surveillance from an unreachable and an unbiased Deck in Space. Its demythologised approach should evoke a rational description of Life Creation. It is a well-stated fact that life exists solely on a planet named "Earth". Thus, it will be easy-peasy for God Point to concentrate more on Earth rather than non-green planets. The non-green planets so far have never shown any clear evidence of life; it may be due to lack of emulating exploration or else, there are no life provisions at all. The cases of dead planets can be ruled out at the moment. The God Point observations are based on random events and obtuse sequences. Thus, the formation of life may be closely connected to the predecessor's events sequentially.

The random spread is the appropriate answer to such queries. To make a concise reply to the query, an event overlapping process must be accessed and unrolled. The day when accessible processes have been studied scientifically, the life questions may have proven answers.

The Life Generation Shaped in a random manner heavily relies on a predecessor step. The predecessors contained optimised energy, which enabled a key generation, which supported further imbricated processes to create a life. Life is an organic format generated with the support of inorganic content in space. The organism needed oxygen from the creation of Earth and its peculiar life-supporting system. *Now going back to the initial question on Life Creation, few lines are presented below.*

To paraphrase,

"Life Creation might have undergone through a hit–miss way with millions of attempts comprising infinite steps of failure and very few interrelated success steps. Life is beyond time, flesh and bones, the imaginative capacity of the brain. Once something is beyond the capacity of brain, the understanding remains hypothetical eternally. Thus, it sounds sensible in either way, Life has created itself or it is created by the supreme, as the person likes to hear…"

Life is indeed an Energy Transition instead of specified creation. Energy had pre-existed in the Universe. The creation of the Universe required an enormous amount of energy to form itself, which explicates energy existence was neither a concern nor disbelief. On a brief note, the pre-existence is the era before the Universe, which is a state of time before a Big Bang. Let us get into further insights about the Life Creation from energy through a subchapter. The basic description is about creation and how this is a part of Energy Transition and spread of energy, which played a dual role of originator and enabler in due course of time.

The conceptual idea of the Energy Transition starts from the following page onwards.

A) **The Energy Transition:**

The Energy Transition and Spread is the Basic activity of the entire Universe, and it is indeed the essentiality of creation. Newton's law of Energy Conservation states, "Energy is pre-existed". In line with that, the Energy Transition of the Universe occurs and as stated in previously, Energy is never stable; it is vibrant right from micro to macro-level. Let us represent the diversification and percolation of Energy right from pre-existence to Human epoch, which is fundamentally a flow representation.

A.1) **The First Significant Energy Transition:**

Pre-existing energy had never been calm and concentrated, although it had hypothetically appeared to be in a composed state. The Pre-

Existence Cosmic Era had been incrementally vibrant and finally had enabled Big Bang. *This was the first-ever significant Energy Transition.* As postulated or proposed, Big Bang was not an instantaneous beginning. Big Bang had a million steps within it, which comprised varied levels of Energy Transitions and respective impacts, which range from micro to macro levels. The elaborative discussion on Big Bang will follow in successive chapters.

A.2) **The Second Significant Energy Transition:**

The formation of the Sun and other stars within respective galaxies became the primitive and primary source of Light Energy for a darker Universe, by an iterative, incremental and incessant process called Fusion. The Sun is the best example of never-ending energy as long as Helium and Hydrogen supplies are there. Their supplies were started since the pre-existence era. The Covert Energy of pre-existence progressively became more vibrant due to imperceptible particles collision of stable energy.

A.3) **The Third Energy Transition:**

The formation of other planets along with Earth by Solar Gravitation. There is a further transition of Light Energy and Heat Energy into Rotating Energy and a pulling/holding Magnetic Energy, which is Solar Gravitation. All the cores of planets are alike and that is why they all rotate around the Sun in a concentric orbit instead of an irregular orbit. The Symmetry and Balance started right from this point. The balancing energy is basically rotational energy. There may be a sceptic on prospective existence or continuance of Sun, stars and the galaxy, but that can be ruled out due to fusion reactions, which never cease unless inhibited of required supplies by any unexpected deviancy eventually or in a jolty manner. The Energy Transition is depicted on the next page for reference.

A.4) The Final Ultima "Life Energy":

Eventually, the Energy Transition spreads over the formation of planets through Magnetic, Electric and Mechanical forms; the Earth was formed along with other known planets in our Solar System. The formation of Earth and its Gravitational Energy ensured positional intactness of all earthy entities. Thus, Gravitational and Rotational Energy play crucial roles in life formation as signified in the Anti-Gravity chapter. Please refer to the Energy Transition flow right from Pre-Existence to Existence as depicted in the next page for an idea.

This illustration is purely conceptual and put for readers' in-depth grasping and interpretation thereof. This should not be correlated to any scientific study or other pertaining studies if any. The picture is developed by me for readers' expediency by visualisation. The energy flow depiction implies that Life Energy is from pre-existence, so it can be rewritten as Life is between Pre-existence and Existence. Thus, Life Journey is nothing but a monumental Energy Transition from Pre-Existence to Existence with infinite duration or immeasurable timelines.

(Contd.)

The God Point

- Pre-Existence Energy
- Compression
- Turbulence
- Vibtating Enegry
- Mechanical Energy
- Big Bang Spread
- Light Enegry
- Heat Energy
- Electrical Energy & Magnetic Energy
- Magnetic Energy
- Solar System
- Earth & Gravitational Energy
- Atmospheric Energy
- Life Energy

> **THOUGHTS FOR READERS' INTERPRETATION**
>
> The lifespan of the Universe may be a billion years; in contrast, Human age may be a maximum of 100 years, who can witness a century and not a billion. If we are trying to leverage our interpretations from century to billion, our brain should intensively work 10 million times than usual as per an empirical equation. Even if the brain is enabled to function by 10 million times, the assumed and inferred thoughts are inadequate to persuade in the absence of demonstrating evidence. Is it not possible to retrospect our thoughts, which are originating from now to billion years ago? Truly speaking, the retrospection is not possible; however, we can attempt to look for some evidence, old aged scripts and a concept laid in this book "a lively imagination" as if we are witnessing era to era from The God Point.

A.5) The Conceptual Triangle of "Life Energy":

The Journey of Life Energy since inception to end is depicted with a Trilateral formation, the Life Energy Triangle. This Life Triangle has several colonies of triangles indicating its transition into various forms within a living organism. Let it be a plant or animal, but the energy spread process is identical among them with negligible variations. Basically,

Life Energy never stops there, as Energy can never be stagnant or motionless in case of animals. Life Energy starts transmission of energy for various life processes and survival functions, which is the inherent nature of Energy. However, the better channelisation will ensure the appropriate energy utilisation and minimisation of wastes.

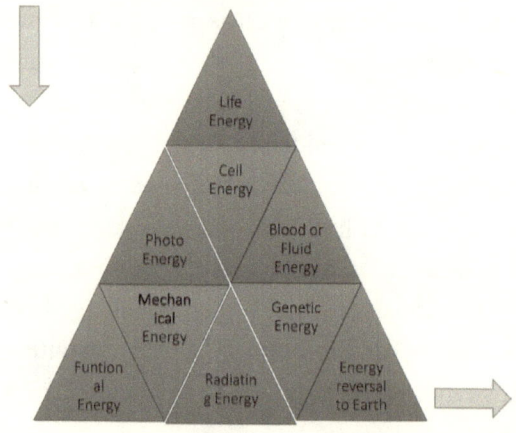

Life Energy has the primary function of sustaining the organism by the exchange of Energy with Earth and atmosphere. The atmosphere is filled with gases especially nitrogen, oxygen and carbon dioxide.

The organism being truly organic utilises organic gases for its energy transformation with the backing of sunlight. Photo Energy is the most important for plant food and thereby linked organisms in the food chain.

The organism is made up of all planetary elements and thus, the evolution and disposal of creatures are from and to the Earth. In case there is a civilisation in R87, that would always be containing an organism, if at all, solely of its domicile planetary virtues.

A.6) **Concept of Energy Transition from the Universe to Human:**

The Life Energy Transition may be synonymous to the religious and popular term "soul". An organism's possession, which is under existence on Earth with a physical format of intricate life process and million seconds Energy Transitions, "The body" when motionless and absolutely zeroed in functionality of containing cells, is mentally, physically and functionally dead and out of vital energy, the energy which had ensured the survival of the organism since the day of inception. Upon understanding the Life Triangle and its sustainable energy flow in different aspects, we can re-emphasise the statement *"The energy and its inevitable random conversion are basic of everything; moreover, it had paved the way for life generation, however, inadvertently"* as mentioned formerly in this chapter.

The Life Creation is an inadvertent, unplanned and random process of Energy Transition and Energy Dissipation. *The Purpose of Life Creation* owes to the energy spread solely as Energy cannot be controlled and stopped. The energy conversion and diversion into various forms as described in the Energy Transition subchapter is the basis of Life Creation. That is the conceptual answer to Creation and its purpose. Therefore, the created entity was the beginning of the phenomenal growth of Organic Matters and growth will continue as long as Life Enablers are reinforcing it.

(Contd.)

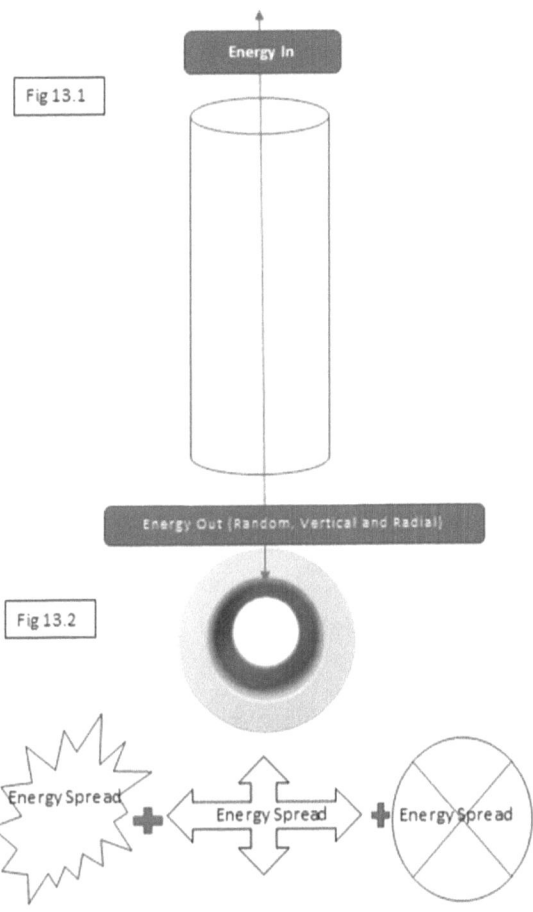

Fig 13.1

Fig 13.2

The foremost step that would have involved in organisms' formation might be Energy spread from Earth and Gravitation followed by Earth's pre-existed Energy forms, which are also enablers, like Atmosphere, Electrical energy, Magnetic Energy and Heat Energy. The Temperature and Pressure are two important components for the life formation backing by Photo Energy.

Let us consider the organism anatomy was a naïve and natural tube as depicted in Fig. 13.1. The organism could have been easy to consider as a circle initially but as it had posed some thickness, so in three dimensions, let us firm the shape as a cylinder, which is same as a tube. The outer shape is shiny and slippery. The circular shape and Symmetry are due to gravitation as explicated in the chapter of Anti-Gravity Effect. The Energy from the surroundings (Universe) might have started flowing into non-living object incessantly for million years and eventually with the support from water and air from the atmosphere, the organic reaction might have kicked off. Thus, there is an instant energy spark of life generation, where an organism initiated its functionalities without any enabler support. This is solely my opinion and readers may correlate with or without pre-existed thought process and understandings.

In the entire formation process, Energy was the commonality, and the Universe was uninterruptedly kept its flow of Energy into organisms before and after. This is because the energy cannot be controlled and must be spread randomly through all available channels without any significant objective. The greater objective was Daedalus Energy Transition, which had inadvertently, randomly and unenticingly optimised itself and generated the Enablers and Life thereof. We are all fortunate enough for such gigantic and incessant Energy support from evolutionary prospects but concurrently unfortunate owing to the inability to cease re-transfer of the possessive energy back to the Universe.

The energy, which is trapped in a creature from the Universe, cannot be kept intact and has to be released or spread on a real-time basis. That is the fundamental reason why we all living organisms are lively and vibrant. The internal energy keeps us moving incessantly, which may be physically, Chemically, Mentally or Functionally.

Refer to Figs 13.1 and 13.2. It can be noticed that Energy within the organism needs a pathway to disburse and going back to Universe or Earth, whatever the reader prefers to take. Primarily it was in shape of a tube, which later modified into various anatomical developments and diversified into millions of creatures in the Earth including the massive plant kingdom. Plant or animal, the basic unit of life is cell, which is a Universal truth. Let us dive deep into cellular structure for further understanding of energy dissipation and spread.

The cell contains a central part, enclosing concentric part between two circular lines and a border for its protection. That is nothing but a thin cylindrical establishment, which later underwent incremental transformation into tissues, organs and organisms as a whole. The primary function of a cell or any living organism is Mobility either physically or functionally or both concurrently. Looking in view of the primary function of mobility, a cell, which has a copious supply of Light Energy and other obligatory energy forms along with enablers like water and air, might have started energy storage and Energy Transition as per requirement. The Energy Transition needs a proper channel within a cell to spread into its parts and also be accessible to all regions. That is why a cell has tubular connectivity within its shape or the entire body. The case is common for both plants and animals. The basic unit of life is cell and we captured its significant functionality briefly.

Organism and Its Energy Connection with the Universe

Organisms were tubular initially for the purpose of energy Transmission and Storage as per requirement.

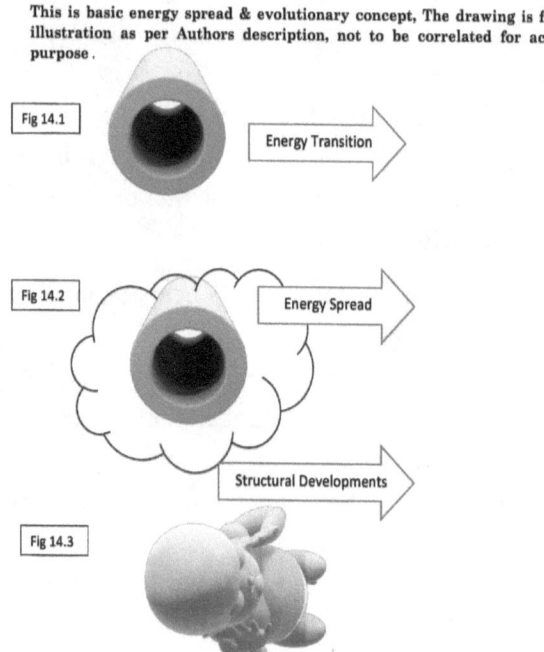

This is basic energy spread & evolutionary concept, The drawing is for only illustration as per Authors description, not to be correlated for academic purpose.

Fig 14.1 — Energy Transition

Fig 14.2 — Energy Spread

Structural Developments

Fig 14.3

Upon going through the presented pictorial illustration, which indicates conceptual transformation idea from tube to organism, we can be certitude on

Energy Transitions routed to the formation of a gigantic, convoluted and revolutionary generation process. It had been more than five billion years since the instigation of such process, which is indefinite until the planet undergoes extinction. The process had enabled the formation of intellectual species of *Homo sapiens*. During the formation of "Homo" species, the process had witnessed significant anatomical transformations such as Energy Transition to holding of energy and Holding to Energy Spread in uniformity over a period through channelising and stabilising. This can be further explicated by taking an example of a non-living entity, which had been exposed to abundance energy supply until its strained destruction or self-destruction. The evolution from tube to a typical Human structure inclusive of internal organs is depicted below, which provides further insights into structuralism of mammals and *Homo sapiens*.

(Contd.)

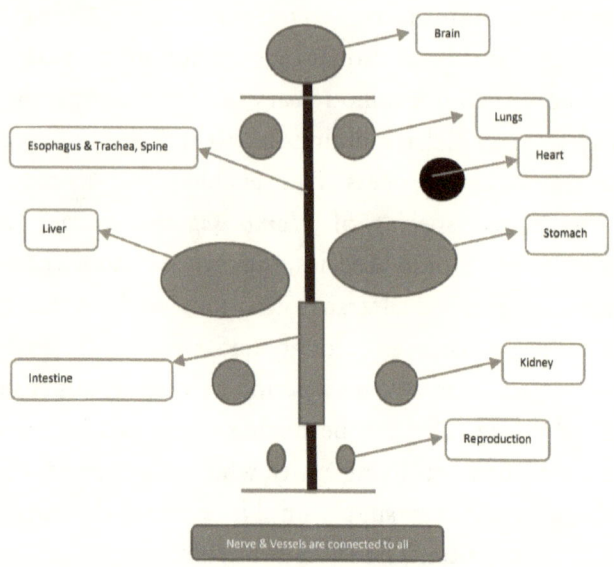

The tube was a concentric circle, which gradually elongated and started intake of food and storage within through adjusted stomach. Eventually, because of the scarcity of food due to extreme conditions or whatsoever reasons might be, the requirement of taking and holding larger amounts of food has evoked. In a similar tone, the intestine need was generated for the majority food absorption by ensuring minimal leftover of food through faecal matter. Thus, the tube twisted and tweaked as per adjusted space within the confined volume.

A Typical Digestive System

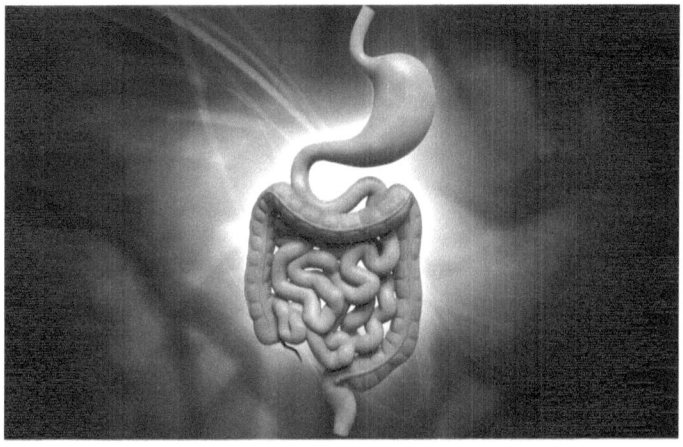

In a similar fashion and generic sense without getting much insight, readers can interpret their understanding as "urge to digest food effectively and detoxification; liver and kidney might have evolved gradually over an enormous timeframe". Time and again, reproduction and all other enablers of organisms evolved. There will be an explicit analysis followed by a general description in "How and Symmetry" chapters more on Human structural and functional format, Symmetry and balancing, which is the basis of Life Creation in the Universe. Forthcoming is about the subdivisions of the Cosmic Energy Transition through funnelling and their inadvertent sequential outcome, which is often re-described as "the wonderful discoveries of, by and within the Universe".

The Brief About Subdivisions of Energy Transition

The Energy Transition comprises six basic steps, which are Energy Passing, Concept of Energy Holding for a very short span of time, Instantaneous Energy Transmission by organism, gradual ability building to store Energy, controlled usage of energy especially for basic survival and life processes and the last and significant one, destruction and reversal of Energy back to the Universe. The entire Energy Transition encapsulated in the aforementioned radical processes and this chain indeed continue for an indefinite period until the enablers and planet as a whole will not support for further liveability. However, the indefinite period should not presume as a very long duration of time; instead, it could be an instant or a short duration as well. In the rarest combination of random iterative Universal bodies and our planet, nothing can be established and confirmed to a closer extent. This is the reality, which is most evident about this random creation.

Now let us reinstate back to the understanding of Energy Transition paces.

(Contd.)

For the sake of better grasp of readers, those six steps can be further alienated into substeps or subdivisions as given below.

1. *Incessant Energy Passing Through the Non-living*
   a) Life generation with immobility
2. *Concept of Energy Holding for a While*
   a) Mobility in living organisms
3. *Instantaneous Energy Utilisation or Application*
4. *Storage of Energy*
5. *Appropriate and Staggered Way of Energy Application*
   a) Energy for Survival, Protection and Growth
   b) Gradual Reduction in Energy Holding Capacity
   c) Inability to store Energy and Utilisation
   d) Destruction (Forced or Natural)
6. *Residual Energy Reversal to the Universe*

In a matter of fact, during such free Energy Transition, the entire process might have encapsulated the above probable intermediate transition paces, which are based on hypothesis. The six stages can be further unified into three basic notions, i.e., the concept of persistent supply of energy, the Storage, Utilisation and Interspersion of Energy and finally, the Energy reversal to the Universe. The Energy goes back to the Universe owing to the process of reduction in Energy Holding and Utilisation Capacity of *The Organism's Organic Possession*, in short, organs and body, which take away the residual unutilised energy gradually through excretion or any available channel back to the Universe irrespective of the type of energy format. The type of energy, which comes into the organism, may never remain the same during dissipation or going out from the system or creature owing to the occurrence of Energy Changeover.

The Changeover process is elaborately elucidated through the pictorial representation of sequential events, which might not be sequential always and instead might be overlapped events as well. The most instinctive idea may come to readers on control factors of such humungous energy spread and transitions. Yes, it is true to state about the encounter with controlling factors; however, even those are inadvertent and random in nature by virtue of their occurrence deprived of any of pre-defined pattern of existence. On a brief note, the visualisation is put forward for conceptual illustration of an idea by/of/in subjective readers and should not be correlated with any emulating scientific enquiry and pertinence thereof. This can be kept in reference for successive chapters' interpretation, implication and extrapolation in general sense or whichever appropriate in reference to readers.

(Please refer to Figs E1 and E2 presented for visualisation.)

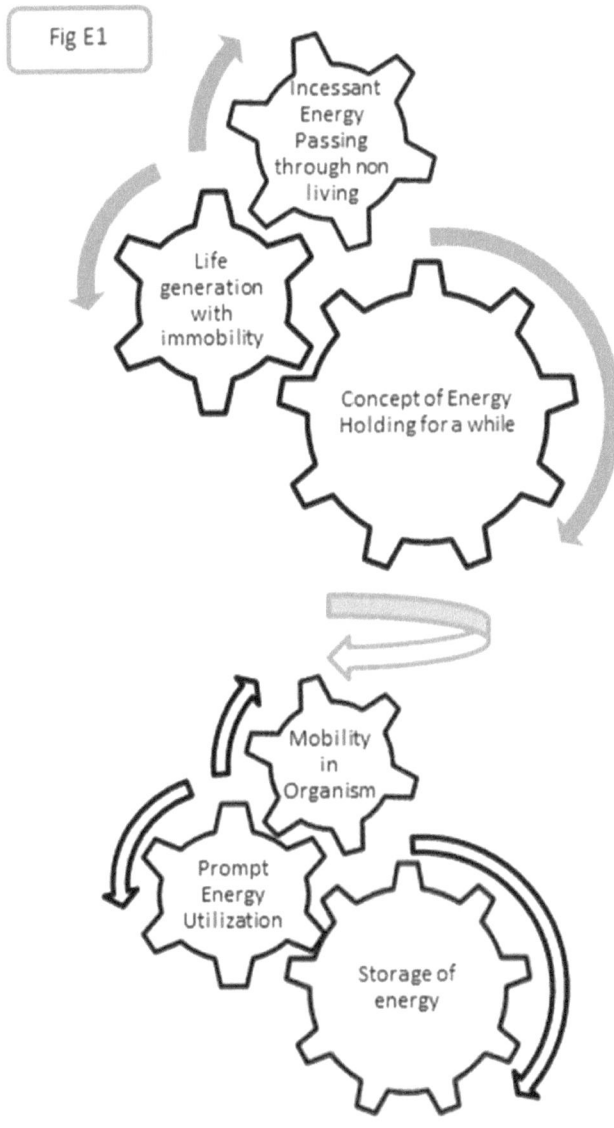

Fig E1

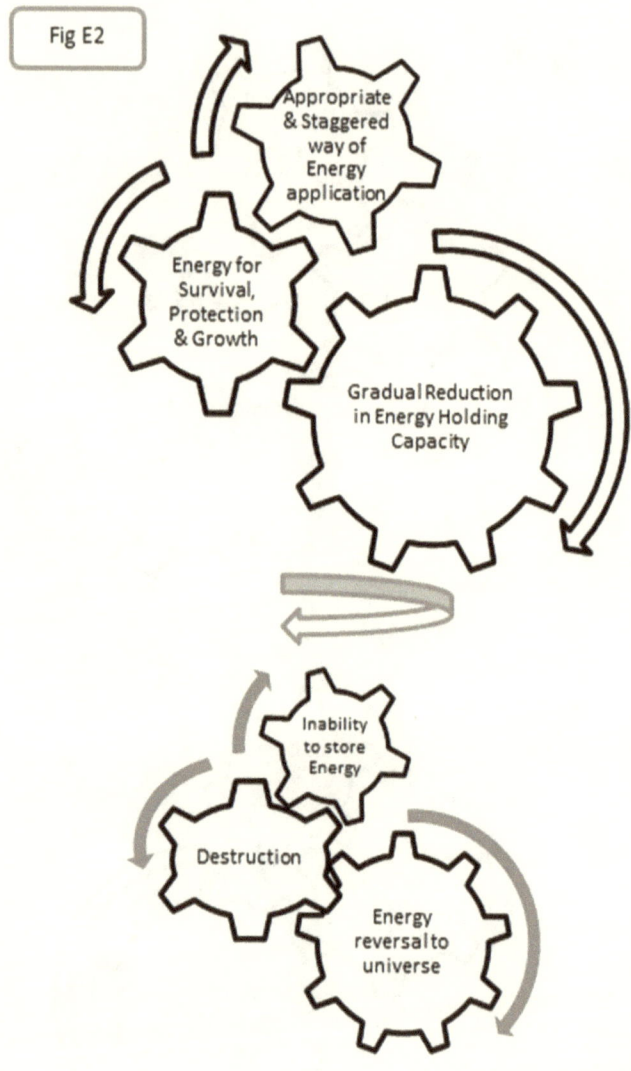

Fig E2

In reference to Figs E1 and E2 from preceding pages, the concept of Energy Transition is virtually cleared; however, in a pragmatic sense, a further need for elucidation may still persist among readers to hone their understandings. In view of this, the depiction of the Human Life System model is presented for further clarification.

The Energy Transition on the Human body from the Universe

The blood circulation system is the immaculate portray of Energy Transition right from the Universe to planet and from planet to Mechanical Activity. The sunlight ensures Photosynthesis in plants. Plants ensure food in parts or whole for Human. Human inhales Oxygen from the planet and then the Energy is transmitted or conveyed through blood for its utilisations on any required mechanical movements required by Organs or Skeletal and Muscular System.

(Contd.)

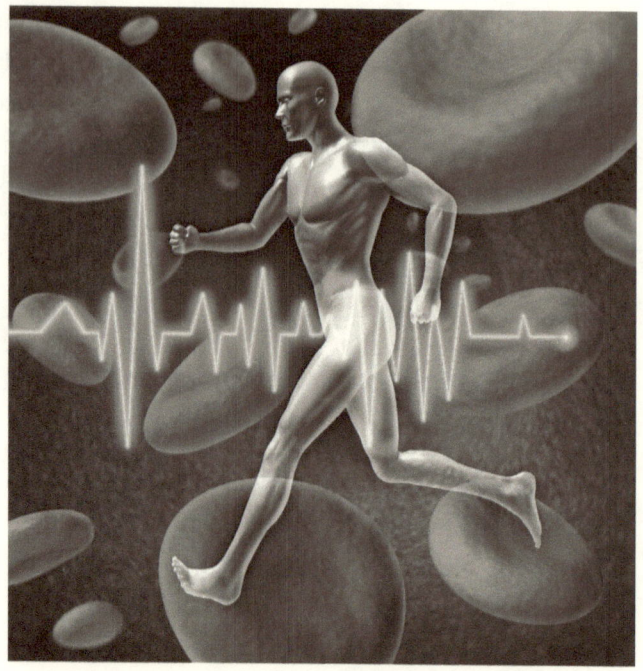

Energy Transition - Solar Energy to mechanical movement through blood

B) **Survival from Nothing**

The pre-existence state of time in the Universe can never be questioned as "nothing" can always pre-exist, which is technically a vacuum or void. The vacuum had enormous and uncountable volume in space in the Pre-Universe Era. Such space has always contained Energy in a very stable form without much distortion and noise like a calm wave. Thus, it can be redefined as a silent and void space with Energy. The Energy always pre-existed, which does not require any divine hand for creation.

The Energy ranged in different types viz. light, wave, heat, temperature, pressure, sound and many other formats. Sometimes I have felt, the Universe started from a black hole, which had huge Spiral Energy whirling its centre. In other words, a black hole is the only pre-existed space in the pre-Universe state of time. Such points are never conclusive because of massive opinion difference among different subjects, theories and religions. This chapter states that Void and Energy pre-existed before the Universe and the pre-existing factors embroiled and headed to Big Bang.

The assumption may be taken into either prospect, first, the divine intervention before the Universe and Big Bang because the divine is pre-existing Energy in religious context. Otherwise, second, the Space and Energy pre-existing eternally in reference to the Law of Energy Conservation. Thus, pre-existence, a perspicuous understanding now, had a key role in the Universe and Life Creation subsequently. Post the Big Bang, the Universe had expanded and paved the way for the formation of Milky Way, the Sun and so on.

The formation process consumed enormous time and energy. The Energy Transitions were uncountable and untraceable because of immeasurable duration, which also contained unlimited iterative transformations. Let us assume that God Point witnessed massive transformations in a very long span of time.

"Why Create Life?"

Let us take the help of God Point to answer this question. I believe that God Point will answer this question without any bias or pre-occupied thoughts. As a primitive step of answering process, The God Point has arranged Life and Universe in order to have precise understanding.

"Life < Enablers < Earth < Solar System < Universe"

When we look at the order, we may feel Life is significant but minuscule in comparison with the Universe. This is a widespread fact and not a new piece of information. The significance here lies with existence from nonexistence. The Life enablers are supportive parameters in the process of Life Creation. The Life enablers are products and by-products of the Earth's creation process.

The tilted Earth's self-rotation around Sun and Earth's natural satellite moon have created an utmost Balance of temperature and light by enabling unconditional liveability on Earth, say terrestrial animals or aquatic or nocturnal creatures. All of them dwell on Earth concurrently owing to existing natural inherent balancing mechanisms, which had taken a million years to establish. Such establishments bridged Pre-Life and Life eras.

To facile previous paragraph interpretation, let us augment previous descending order and represent in below fashion:

"Existence < Life < Enablers < Earth < Solar System < Universe < Pre-Existence"

If we take out all enablers in between, we can get an order as below.

"Existence < Life < Enablers < Earth < Solar System < Universe < Pre-Existence"

Now it is explicit that *Life is between Existence and Pre-Existence.*

"Existence< Life < Pre-Existence"

Let us decode this by further insights:

Before creating life, we need stable and suitable parameters for such miraculous creation. Such parameters collectively exist only in one planet "so far discovered", i.e., Earth. To create life, enablers are required, which are other than planets, such as Energy, Temperature, Light, Pressure, Air, etc.

The God Point further elucidates that life was an organic transformation by imperative steps until perfection had attained. This does not mean that evolution stops here, it continues indeed. A Human may or may not exist post a million years later if we scale up to 3–5 K years from now; we can see a massive transformation in Humans and the "planet as a whole" through an ongoing evolution process. The God Point feels that Human is not a perfect creature as per current structure owing to Symmetry; Humans should further strengthen their adaptability in line with ongoing and future changes in the planet.

The shapes pertaining to *Homo sapiens* need not be transformational, but the evolution pertaining to the functional side still needs to be enhanced than current standardised capabilities. For example, The Capability to Listen, See, Think and Utilise brain can be strengthened beyond adequacy. From the previous paragraph, we can derive that the spread of randomness on Universe creation and expansion is prevalent and thus, *there is no purpose of Life Creation. The purpose would never be created before the entity is produced.* The purpose can be described in a generic sense but not in a particular pre-designed pathway. To understand the above lines in a better sense, please refer to the bpm level comparison in Chapter 1.

"Any entity undertaking zillions of activities in its lifetime randomly under billions of frames (a view from The God Point) can never be predicated frame to frame."

I mean the frame is a state with respect to time in the above lines. Envisaging such unpredictable scenario, Life cannot be created with some pre-defined purpose by the Supreme. For example (in an implied version) for achieving a milestone of 10, one must start from one.

Therefore, one and ten must be pre-defined as per life purpose fulfilment. What happens in actuality is as depicted below.

Plan: 1-2-3-4-5-6-7-8-9-10

Actual: 1-6-4-2-3-5-7-8-9-10 and other possible actual ways as:

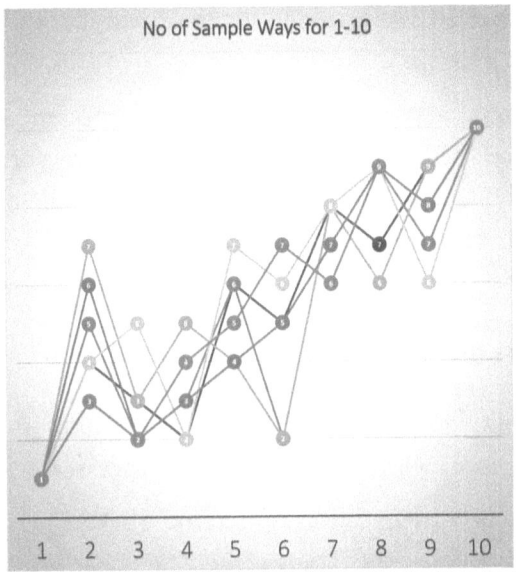

Actuals can be numerous apart from the above, which need not be elaborated as readers can guess. The above sequence can never be predicted when "n" number of combinations can be possibly encountered between two significant events, the so-called life purposes. Thus, this simple example can easily conclude that Life cannot be created or generated by envisaging a purpose.

Actuality may vary in many other ways as long as there is no mandatory requirement for linked activities.

Such linked activities are cited, especially in two cases, one is voluntarily, and the other is involuntarily performed activities. Involuntary cases can be mostly life processes that are biological sequences, which must follow an order. In contrast, the former category is voluntary, which must follow a sequence command purposely under an instruction like encoded mode or wilfully, which is a habitual approach.

Although predictability of the aforementioned biological sequences is high, it is never controlled naturally unless inhibited by any external aid. To determine predictability, let us enlist typical reactions for a random action.

React or Do not React

1) Instruction – Follow or Do not follow
2) Involuntary Process – Inevitable processes
3) Reaction with respect to External Factors
4) Instinctive and Spontaneous reactions
5) Reaction influenced by experience
6) Shock, Grief, Fun, etc. (all expressiveness)
7) Mental reactions without physical expression
8) REM/Dream is the reaction of the subconscious state.

What makes the next step is the previous step; the previous steps are the result of infinite sequences. Thus, life events are never predictable at the bpm level, and life has no pre-fabricated purpose, but yes, the purpose can always be there, unless that Human life is insignificant in comparison with other mammals. Despite having a recurrent reactive pattern, the next reactive moment prediction can never be superimposing with actual reaction.

This answers the "purpose question". Indeed, the answers to *"If somehow life was created, does this creation deserve any purpose?"* may look insatiable and arcane to many people.

Summary

The Life Creation is an inadvertent, unplanned and random process of Energy Transition and Energy Dissipation. *The Purpose of Life Creation* owes to the energy spread solely as Energy cannot be controlled and stopped. The energy conversion and diversion into various forms as described in Energy Transition subchapter is the basis of Life Creation. That is the conceptual answer to Creation and its Purpose. Therefore, the created entity was an endowment by the Universe and of course a great beginning of phenomenal growth of Organic Matters. The growth will continue as long as Life enablers are reinforcing it. In this subchapter of WHY, readers were explicitly dived through a concept of "Why create life" and "Why life had formed and was expected its formation encompassed a purpose within".

This WHY subchapter stated life purpose and creation mechanism right from the Big Bang or pre-existence era of cosmos. The purpose was basically an inadvertent Energy Transition, which is further subdivided in this subchapter with a real-life example and appropriate pictorial illustration for readers' convenience. The conjectures stated by me are unique. The widespread opinions were never questioned, and readers may keep their take away coherently. With this note, we move on to the next subchapter of "HOW"…

HOW

The subchapter HOW, a conceptual approach of creation put into, is solely based on my opinion and readers to follow in line with their thoughts and take away the parts or as a whole.

(Contd.)

Are Life Imbrications Being Altered?

The expression of a distensible format is, *"Can we alter the imbrication of life events of day on day, can any event be either revisited without adjoining and distorting entire chain by hop-skipping intermediate events."*

The interpretation is to cognize about the alteration of the past or future at Human convenience. We may feel the above question ingeminates like an age-old query about time travel although it is not. This is about the avoidance of life events overlapping or memory overlapping.

This is about bringing back the experience and feeling those "as it is" which is indeed coinciding impeccably with "as it was", let that be rarest like the first kiss or first touch of romantic subjects like lover, fiancé or wife. Let that be a bad or good or neutral by nature at that point of time.

"In another way, this concept is time travel of feelings, memories and emotions Instead of Time or Events in true sense." "This may not be confused with déjà vu, which is an experience of being felt or witnessed already for a look-alike situation for an ongoing event."

Such concepts might have been thought by many individuals in their life journey, certainly, but never attempted for retaining or re-experiencing again due to a forward approach within them. The forward approach makes their retrospective nature unnoticeable to themselves. By the time folks realise the best days "so-called sweet days or harmonious time", they would have crossed even present. This happens in due course of embracing themselves to face artefacts turbulence and pseudo-criticality created by Human in the pursuit of being happy in the Happiness Chain.

A brief about Happiness Chain

Happiness Chain is a never-ceasing and never gratifying chain among similar minded people. People find their happiness by making other people happy who are in linked connections of such intricate chain. In fact, those people search for self-happiness upon attaining milestones planned by others. As a part of the reward, this is ok but searching for Absolute Happiness in such mechanism is foolish expectancy.

Concept of Happiness Chain

How the Imbrication affects whole life and Why this should not be perplexed in disguised ways instead of a desirable direction?

The intangible parameters surrounded by us are bound for conversion into tangible ones at some point in time. The time is, of course, unspecified and limitless; nevertheless, it is no matter of concern for our further elaboration. The imbrication of events in one's life is incremental by nature. Such imbrications are perpetually retained forever, and brain seldom, indeed never, forgets any sequence. However, the absolute reclamation always remains a concern, at least, on average. Let us take a basic example to make this analogy a paraspinal one as an illustration attempt.

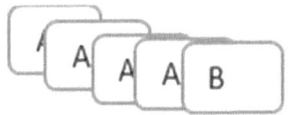

Upon looking at the above picture, readers can guess five sequence events, and four of them are recurring. This is what a generalisation is. However, upon a closer notice and vigilance, this has imbrications, which eventually created or enabled event B. Thus, ***the imbrication can never be neglected*** once in reference to the below picture, which is an expansive and explicit format of the chronology of the above events. In fact, there are no recurrences and all events are precisely unmatching with each other; thus, overlaying of them evolved a state of time of occurrence of B.

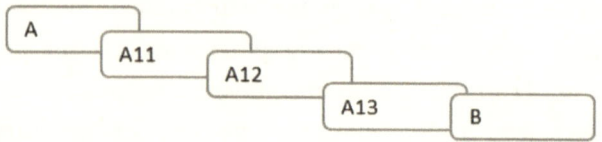

If disguising of events is not decoded to the factual state, then a perplexing state of mind evolves into the picture. The quest for precise pre-action will be bombarded across the mind cognizance borders but no solution will stay bidirectional because of a confounded input event or sequences if plural. However, if the action was taken for A 11 and A12 on the basis of A, then A 11 may result A 16 or A-16 or whatsoever, which is not predictable at all.

Every event has the absolute and relative reality, which ought to be an unmasked and appropriate decision to be taken for a desirable event as resultant instead of perplexing across the way and finally leading to time mismanagement. This is how the How works in life, and thus the life imbrication plays a significant role.

Conclusion of "HOW"

"The question here is do we really need to worry about the past events and future events which are unconditionally beyond our line of control? Obviously, a Big No. The fact remains a common answer, which is the state of time that needs to be controlled rather than optimised with a basic level of predictability."

and

"The integrated happiness within the self is never attainable due to infinite nature of the connection and perception chain, which is forwarding by imbrication and without a defined ultima. The happiness within the self will be achievable once people detach and refrain themselves from such intricate chain, the so-called "Happiness Chain", and define their ultima of life within their lifespan. As stated previously, the emotions are created biologically and chemically by and within the *Homo sapiens* by aggravating and imbibing themselves with the pseudo-comparison of the outside world either to prove supremacy or avoidance of inferiority in a relative manner. The integrated happiness is a resultant of absoluteness of one's thoughts, which are free from social relativity."

(Contd.)

The "How" Extended

Energy Transition Within Contemporary Human

The Human's Organic Possession needs energy for everything right from its inception within the placenta. The necessity of placenta is to protect and ensure growth until the adequacy level where the baby can survive by itself self-reliably. Once the baby comes to the planet by stroking its soil and encountering its gravity by itself, it starts everything of its own for survival and life processes execution. The activities may involve, but are not limited to, Absorption of food energy, energy generation and storing and utilisation of energy. The utilisation can be divided broadly into two classes, Energy for Life Processes and Energy for any activities pertaining to energy spending or calorie expending.

Let us understand what the basic form of Human is and where it starts. The foremost basic of everything in the Universe is the state of matter. What does nature comprise? The organism's formations will be neither out of deterrence to nature nor away from planet properties. This is the basic element to be re-emphasised for the understanding of radically dribbling process in the process of Life Stabilisation on the planet. The organism of the Earth is purely made up of planet Elements. Therefore, we will not find any alien components in our body. This is so radical and significant distinction about our creation, which indicates the best utilisation of planetary resources for a systemic formation, which simply ensures the random purpose of Energy Transition and this process is incessant and divergent due to creation of innumerable species on the planet. However, those species might be extinct, but their transformation spreads or genes still might have been continuing or ongoing even in the contemporary era. In continuation to the previous context of earthly virtues and their relationship with Human structure (in and out), let us evaluate the coherent pattern of Human genesis evolution.

When we cogitate about Human structure, firstly mind points to water. Yes, water subjugated the Human, with 70% of its share. However, this may be an inevitably vital and ultra-essential component but not the only one. The second dominance of the body is the Skeletal System, Muscular System and Epidermal Areas as a whole. This is merely shaping the structure and movement. Obvious to restate that the Energy Transitions major components and POAs (Points of Action). The third dominance sprawl the internal organs, which are broadly the tissues, cells, blood vessels and intestinal organs having vital functions as compared to first and second superseding factors. The fourth most and significant factor is the connections and their flow or the mobility within Human to connect each and every cell, tissues, Muscular System and Epidermal areas and ensuring the intactness of the adequacy meeting requirement and serving to the whole organism. *Radically, the flow and connections ensured the Integrity of an Organism. The Organism is immobile, of course, Lifeless without the flow. There are prioritised flows, which are highly crucial and their milli of Time dimension (Seconds) will encounter the probability of liveability.*

Thus, the four components are the rudimentary and significant aspects, let that be Human or Any animal, of creation and survival. Even plants also contain a majority of four components. The Organism extracted from the Planet and a completely sole composite matter evolved over a substantiate long span of time. The Organism is Technically a mobile mixture of Water, Gases and Organic elements from the planet. The composition within a Human, of course, is a whole sole planet property, which everyone aggresses. The human has no mobility or life unless an Energy Transits into it to make self-reliant to lead and earn liveability. Even if we are fully created by the esteemed planet: The Earth, we still need cosmic energy or its descendent for being alive. This evokes a question on the existence of Alien or Extraterrestrial living beings as planet and Universe are still common. However, the Planet may be very different from our Earth or else possible the same to same, as the Energy Transition generator is common, the Universe. Thus, the Life Energy is common between extraterrestrial and earthman, as they pulled it from the same Universe. The variation could be in absolute differentiated state and pole apart in terms of Physical and Chemical Compositions. This could be any extra periodic table gases and water equivalent liquid availability, as such an unknown planet could make us certain about existence. Now let us get back to the composition of Human, as mentioned earlier, the four components. The four components thus may be common for that planet as well, let us say R87, for example, the planet name. The R87 must pose four necessary components for its planetary living organism's creation. Birth and Survival are in an intricate cyclic pattern. The cyclic pattern may have fewer or higher intermittent steps but certainly, the commonalities in the birth to death process cannot be ruled out in either case. Thus, Life Cycle must be a common function in the Universe.

The Celestial Culture and Formations

To devise imagination and adjugate subsequently, we must understand basic anatomical structure and composition of any planet for almost all living and non-living creatures of such planet. Upon restating the previously mentioned concept of four components, every planet must sprinkle all its unique components for its composite organism.

The planet is tabulated as below

Liquids/Fluids	Water replacement
Muscular and Skeletal System	Calcium and Organic replacement
Internal Organs	Depends on evolutionary Pattern from a tube
The connections between the whole body	Electric or light or sound, whichever energy can travel fast and be available on such planet

Gases will vary from planet to planet; evolution to vary from planets' inherent conditions; temperature and pressure suitability may vary significantly.

R87 > Earth – More evolved Humans

Life Age of Earth > R87 – Little laggard

All these depend solely on Planetary Individual virtues exhibition on Life Generation and Growth thereof. Therefore, the existence of pre-existed aliens, Existing Aliens and Advance Aliens than Earthians cannot be ruled out. Thus, Life can be generated anywhere in the space with unique kind of its format, shape and sizes, but we can see or foresee aliens without any doubt. Is the Creator or God Same for Both Earthian and Alien? Of course, The Same Universe is the Creator through inadvertent, Enormous Hit and Miss attempts of Energy Transitions, which are perpetual.

The planet has endowed many astounding things eventually that are the reason why planetary living creatures are relevant and ingenuine to their sole planet. For example, the height of an organism is fixed by nature. Nature has plants and other inhabitants; all have certain height limit owing to be contingent with the surrounding and other creatures. If that varies, it is quite natural to see a gigantic or dwarf Human on any planet supporting life conditions. Human Should not be misinterpreted as the best evolved living creatures in the Universe; there may be some extraordinary extraterrestrial format, which has advanced mechanism evolved within peripheral of their individual planetary virtue. Such advanced evolution might be uppermost with respect to functional, structural and visual as compared to Human and this may look weird or wow to a Human when it is encountered. The emotional constituents like hormonal, intelligence and behaviour pattern may be pole apart or even very similar. The purpose to explain this is: Every planet has unique properties to create a unique aspect of the organism by their structural and functional pattern. A planet may not be able to generate life even or some planets might not sustain life albeit they can generate life. The human evolution thus is not a big surprise and divine driven, this is just a result of million hit–miss trials and random events sparked through one in billion unsparked eventualities.

(Contd.)

The God Point

"THE HOW – THE DESIGN OF Human"
and
"THE HOW – THE FLOW IN HUMAN"

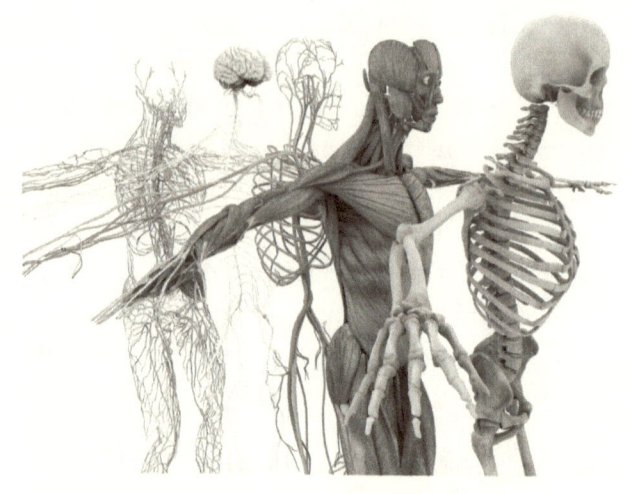

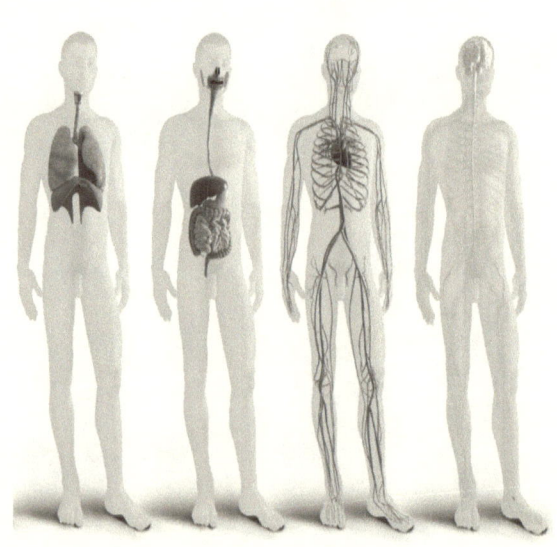

The Specialism and Uniqueness of Human Design

The basic function of Human is Survival and Mobility thereof, the Mobility of Organs, Connections like blood and so on, Limbs and Skeletal System and almost all cells within the body. As long as mobility is enacted by the organism, its existence prevails profoundly else the status to change. The uniqueness comes through systemic and intelligent inheritance of the Human body. To have further insights, let us dive deep into a few significant aspects of life. The cell evolved as the first and foremost mobile unit. The structure of the cell is merely circular with a central part comprising little surface area.

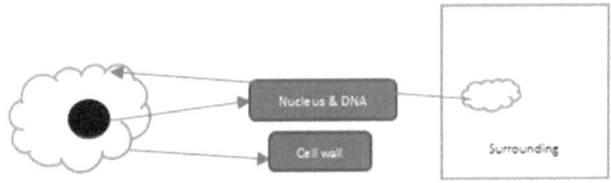

As shown in the picture, this is the first available structural organism on Earth. That is indeed a simple design with an irregular border and a centre with some minerals and surrounded by fluid, the cytoplasm. What could be imagined beyond such a simpler structural formation? In my opinion, a cell or enclosure is from an open and infinite surrounding, without nucleus/the central part. However, if the central part expands itself on the structural front, it has to follow a radial pattern from the top view. However, in three dimensions, it must pose a certain amount of thickness, in line with that, it can be refined in the context of expression and redefined as a tubular pattern. The tubular structure thus expands into a tubular pattern, maybe irregular geometrically, owing to the uninterrupted supply of energy as described in previous chapters. Upon passing through various intensities of energy forms, a striking phase of energy supply may have strengthened eventually.

Subsequently, the strengthened striking phase of energy might have triggered the potential difference to a stage, where electron and protons mobility instigation is certain. Thus, there is spark generation. In other words, the mobility within an organism is generated, which is Life and obviously a life inside a Basic Organism. This could be further represented pictorially in subsequent section as per Fig No. 14.1, 14.2 & 14.3 Page No. 74 The foremost structure is an irregular tube.

The structural formation gradually expands its horizons minute on minute, and this is because persistent energy flows into the organism. *The Cellular Organism has started absorbing all energy instead of a cent per cent passing throughout it, however, gradually and very slowly*. The process of absorbing energy had cascaded to the energy utilisation process but the utilisation at the beginning amounts to decimal while the energy absorption might be a whole number. *Eventually, the process of holding energy and disposal of residual energy is inculcated and developed by organisms*. Once the process of holding energy started, the cell needs some micro organelle within for storage and regulation. Eventually, due to subtle potential difference between two portions within the cytoplasm, a bridge is created and energy is transmitted within whenever there is a copious supply of energy. The bridge is Mitochondria, the powerhouse of the cell. Once the cell has attained the capacity to store the energy, concurrently it might have begun Utilisation for its Growth until the Growth is inhibited and pained by external factors. In the process of Energy application, the eventual growth demand necessitated further enablers within a cell-like reticulum's nucleus and incremental energy absorption. As stated earlier, the cell which has tremendous potential of unwinding further opportunity must be commanded to be in a structural manner either for deterrence in any energy mismanagement or contrarily for vigour underutilisation. Therein lies the intriguing concepts of gene, which is positioned at the core of cell in general literary description or within "the nucleus" as per biological terminology.

Please be noted that genes were encompassed by multi-protective borders. Those protective membranes or walls positioned themselves right from the cell wall (in plants) or else cell border (in the case of animals), which is distinctive with respect to the type of organism, to nucleus's wall and another cell organelle in between. Hence, the concept of protectiveness or self-rampart idea of any organism evokes for further elaborative elucidation, which is covered in consecutive chapters.

The location of the nucleus was, of course, an inadvertently strategic location because of operating and controlling from a single safest point. The Energy, which is under holding or stored condition within a cell, might have started dissipating gradually for life processes right from growth to organelle formations. Undoubtedly, this might have undergone many years of infinite transitions and transformations. Eventually, entire cell organelles had formed, and a nucleus had positioned within and at the centre of a cell as stated previously. *The nucleus was indeed a core of the cell, which comprised Chromosomes and DNA, a unique aspect of the cell, which keeps miles away from posing any similarities with any other cell within the cosmos. The entire cell will be instructed by gene for all end-to-end activities for energy optimisation and regeneration.*

After going through the above passage, the notion of Regeneration and Reproduction is to be evaluated explicitly in all possible considerate directions. Obviously, that must start with intruding queries as spotted below and have blurred in the readers' mind, may be or not, at this moment.

On prima facie, my conjecture should not be emulated and correlated in a purely academic sense. However, the conjectures are generalised and abstracted in view of general understanding of evolution and with respect to variable Human behaviour.

The Two Questions

"What made the reproduction to happen and is it spontaneous or forced by any external factors?"

and

"Are regeneration or repairing required and how the cell started restoration by and for itself?"

The First One

The organism has a peculiar virtue of urge. The urgency always directed to attain the desired objective no matter whatsoever way it follows, this no way constraint to energy spreads unless voluntarily controlled or optimised or interfered. What drives the organism in a pragmatic sense? The organism is driven by its sole purpose of survival; any other point beyond this is coherent and enablers.

The cell, which started storage of energy and its energy utilisation for its growth, eventually develops micro organelle within its frontier. The organelles were formed, not instantly, gradually in infinite successive steps as stated earlier. Indeed, the micro organelle of the cell had no role initially except for survival and growth support. There in the external environment, intervention might have instigated. Not all the surrounding external factors were enablers, few were destroyers as well. There comes the concept of self-defence. In the process of endurance fitment with surrounding, the cell has created a potential difference and a barrier as a protective measure.

However, the question of self-defence never arises unless neurological development was self-demanded and formed in the initial phase.

(Contd.)

To understand this, let us revisit the Energy Transition episode:

1) Energy Intake
2) Instantaneous Mobility Unwillingly and Uncontrolled, Vector-less and conclusively, without any coherent purpose
3) Holding Energy
4) Mobility
5) Uncontrolled Mobility
6) Controlled Mobility within an organism

The neurological sense of organisms might have evolved after mobility to control mobility direction. This is again a concept of Energy Transition in a controlled mode. The urge to React or Respond is a sensible crucial factor, which catalysed the electron movement and nerve coordination thereof. An initial cell, which contained basic components of Cell Wall, Nucleus and Cytoplasm, had pre-existed potential difference and nutrient might have instigated the flow of Charged Particles like protons and electrons. There is an explicit analysis about *the flow* in the subchapter flow; hence this point can be hop-skipped, and the appropriate reasoning for the question can be conjectured and conceptualised. The reproduction is primarily an asexual process at the very commencement stage of cell evolution. The purpose of asexual reproduction is to cut/shed a part of it to generate a similar one to parent, eventually after desirable growth and basic anatomical developments. The case is even applicable for self-pollination where both the variety of genitals are unified in the same organism, but eventually, on demand of genetical variations and ability to survive under adverse or relatively extreme conditions, the relativeness varies era to era.

The cross-pollination in plants or cross-breeding in animals might have instigated through a media or self-instinct. This is obviously a widespread and common answer, and there is unbendingly no freshness in this topic. My purpose is not intended to a generalised understanding, rather a deeper insight, vibrant intuitions and a coherent imagination, which is proximal to reality or can be effortlessly simulated to a typical real-life scenario. Therefore, I opted for a view from The God Point on answering the prevalent questions.

A View from The God Point on reproduction and regeneration: the urge to reproduce eventually became a strategic, coercive and involuntary decision by the sovereign organism, the nucleus of a cell and brain. This could be abridged from an intricate median process by revisiting to the beginning of the first ten basic unicellular organisms. Putting our eye to eye in God Point, we can see at a very nascent or budding stage, the cell division is a reproduction for a Unicellular Organism. The Unicellular Organism has a typical structure of border, central part and cytoplasm, which comprises mobility particles such as fluids, electrons and protons and of course water and nutrients for a broader purpose of conduction and survival ultimately. The basic cell continued its growth until a saturation phase. Eventually, owing to the crossover of the threshold of potential difference, which was pre-existed in the cell, there was Balance and neutralisation was necessitated for the cellular body for maintaining survivability. In the process of balancing the overall cell, the mobile particles realigned themselves in a grouping pattern, which are at the either side of the cell as per physical description, as the difference has gradually expanded with their unique identification, while the individual group protection mechanism enacted in place by both the regions of the cell, thus, ensuring a protective border development.

The cell division was enabled by the subtle potential difference, the potential difference in true sense not only applicable to an electrical mechanism but also, it can be a difference of anything wherever Energy Transition occurs. The aforementioned difference might have instigated the cell division asexually and which later started a multicellular configuration of colonies or tissues, which has a greater role as teamwork instead of individual potential. The urge to reproduce in a cell perhaps instigated with the self-balancing mechanism to avoid self-destruction because of excessive growth and equilibrium of the whole cell.

The central part, the nucleus, of course, contains basic elements along with a command of pre-defined growth and divisibility pattern.

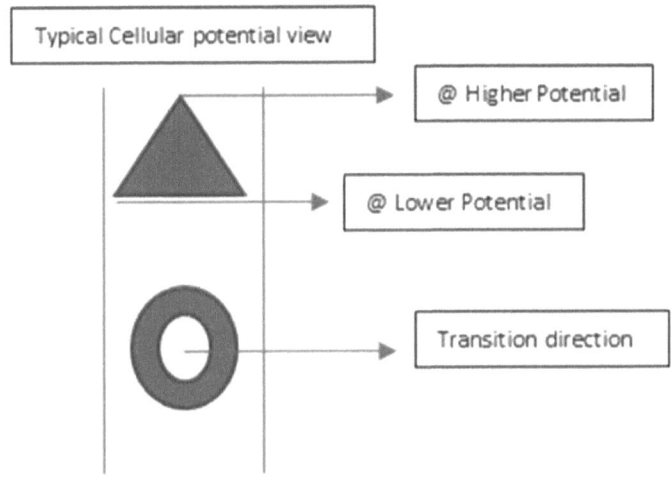

The nucleus is at the central part of a cell, and the astounding part is, as indicated earlier, "The central positioning", best sounds in the purview of protection, strategic, and coherence. As depicted above, the transition direction starts from the centre or mid-point, due to concentration of energy or higher energy density at the initial phase, to the outer area of the cell as it expands thoroughly in a systemic manner by delivering the energy peripherally.

Let us extrapolate the cellular potential to Human as a whole instead of individual veils, where everything starting from the brain and body electrifies, which initiated heartbeat as well. The brain is central and at the highest potential of the body, for the sole function to control and regulate the whole body just like the Triangle as depicted in the figure. The brain ensures all intact connections and responsiveness of the body. In fact, the Human body is an intelligent energy generator and conductor of energy to all POA as described previously. The flowing picture demonstrates that the intricate and vacuolate functions reach the Human Brain and there is a potential transfer to the whole body as per the energy requirement pattern. After going through all above, it is never weaned to say.

(Contd.)

Sympathetic System

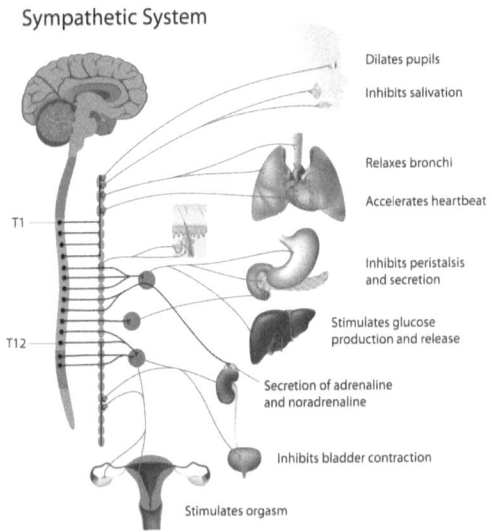

Parasympathetic System

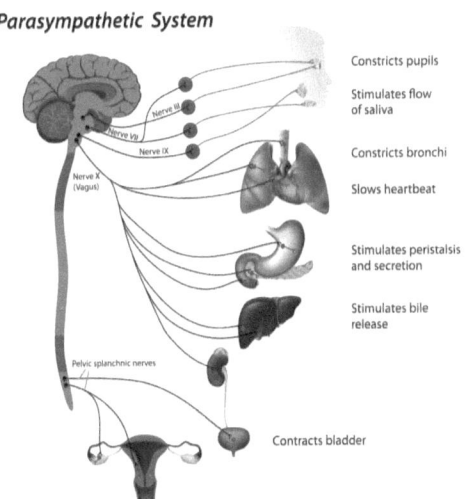

The Human is one of "Visible format" of Electrical Energy, which has taken a physical shape and has a capacity to grow, regenerate, protect, cleanse, response and self-destruct.

This energy has continued interaction with the environment and certainly does affect and gets affected by the so-called external factors and even extraterrestrial factors as well. The brain has the capability of managing the control and feedback mechanism.

This is possible because of positioning at higher potential and ability to generate electricity, which travels through for Energy Transition: a radical purpose of Human creation and ongoing evolution. Now, after elucidation through energy approach (albeit redundant) and unwinding hypothesis of potential difference, let us get back to the question on the Urge to Reproduce… As explicated through the line of potential difference, the reproduction through budding or a non-crossover breaking mechanism is owing to a self-balancing mechanism by equating potential difference to a desired level for long-term liveability. Then comes the concept of the potential transitions from high to low in the process of energy spread. During the transition, the cellular transformation from "unit to multi" and penultimate the ultra-evolutionary concept of brain comes into pedagogy. However, the first Human alike mammal, somehow undergone an infinite transformation from any multicellular organism, is bisexual or unisexual, the evidence to confirm.

However, as per imaginary God Point, there was no reproduction mechanism initially, and the first mammal was basically FEMALE, NOT MALE. The women creature had womb to produce offsprings and mammary glands for feeding, even today men still have nipples, which they really do not need. At the nascent stage, there might have been asexual mechanism, genital developments and gradually cross-breeding of interspecies, then intraspecies and strive for perfection in a line of Survival of the Fittest. Eventually, mate finding was an activity, through arousal by genitals and ultimately, the brain, which has self-constructed hormonal chemicals for triggering instructions as per the external interaction level and suitability. That is the reason, interspecies eternally stopped and the same species enamoured to it. This hormone triggers the need and the hormonal mechanism or The Urge, and the brain triggers that. The Urge to Reproduce is a sole purpose: Balance the energy and regeneration. All these assumptions will channelise our scattered understanding of the Universe and life into a definite approach. Now let us revisit one of the ever-important and significant question followed by the answers of the previous question, for an illustration basement.

(Contd.)

THE HOW- THE FLOW IN Human

The Flow Has Its Own Importance.

The typical flow selection strides are put forward as follows: *Radical Flow choices or selection available within an organism*

1. Solid flow option
2. Liquid flow option
3. Electric flow option
4. Gaseous flow option

Now let us be perspicuous on the understanding of all flow choices and their true coherence with respect to the choice opted for. The major distinction between transition and flow is the state of matter of the dwelling surrounding or environment.

First, the Solid Flow Option

This choice requires massive physical thrust within and by the body. The plausible purpose of such flow is the intake of food and disposal of solid wastes from the body. The solid flow typically springs from the top to bottom through a cavity, e.g., end-to-end digestive and excretion systems, unless there is Anti-Gravity, e.g., regurgitation or any similar obligatory case.

This solid flow requires a through the cavity and thus is common for almost all creatures, say vertebrates or invertebrates. The absorption is the basic requirement of solid flow management.

Second, the Liquid Flow Option

As it is very often regarded that *Homo sapiens* consist of 70% of water, the liquid flow is the most predominant and significant flow in any Organism and Human Body, of course, is not an exception. The liquid flow mechanism is through a flexible bridge of potential difference. There are four crucial mechanisms integrated into liquid flow viz osmosis, pressurised flow, reverse osmosis and natural flow due to gravity in peculiar cases.

To understand the flow, readers may correlate simply with Human Structure or a Simple Tubular organism, whatever they feel is convenient. Let us introduce POA, which requires liquid flow.

Radically the purpose of commuting a liquid is to carry required elements to POA from any part of the body or designated part of the body. This is a very natural phenomenon indeed, which requires a natural pumping mechanism for travelling mechanically from one point to another. Wherever the pumping mechanism is not available, the liquid flow mechanism will transform to osmosis and if not, modestly a free fall i.e., gravity support from nature.

This flow is a self-same natural process, irresistible and embarked by required elements for POA; however, any interventions can certainly hinder. This liquid flow is even a part of a gigantic Energy Transition concept as explicated in earlier chapters. The energy what we receive, store and spend is an unconditional descendent from the Universe, and it spreads and eventually disperses back to the Universe maybe today or some other day, however, in strict adherence of Energy Transition principle. In a matter of fact, there is no such principle to be followed over command, but this is an absolutely natural process established as a principle. The most important flow in Human is shaped by liquid mode especially the blood flow, which carries the penultimate stored form of energy. Supposedly, any part of an animal body required any movement (maybe triggered either voluntarily or involuntarily). The blood flow and the potential difference fulfil the requirement. The movement is mechanical Energy Transition in nature, which ultimately should go back to the Universe either directly through POA or through few penultimate steps before whole dissipation or final Transition. The energy thus propelled out from the body and wholly transmitted to the Object of the Universe, through POA of any subject (the living being) by making the body to remain deprived of energy or say exhausted. Not only that, the liquid flow is a massive mode of Energy Transition from any living creature to Universe or vice versa, but also, it is a saviour of whole living being or organism. Once the liquid flow is interjected and its re-instigation seems biologically infeasible, the liveability of the affected organism is said to be at stake and the organism transcends from "is to was" and remains as a memory of an ex-inhabitant of the planet. Therefore, the flow is supercritical for survival and Energy Transition as well. Thus, any flow can be called as Life Energy and as soon as the flow stops, the Life Energy is diffused back to the Universe, termed soul in religious terminology, which we are going to encounter in upcoming chapters.

The Liquid flow option is the fastest available option of Tangible Energy Transition to all parts of an organism for any life process in the form of liquid and gas within the organism, and that is why organisms selected this mode for mainstream utility functions within them. The coherent reason for this mode is plenty of supply of water from nature.

Third, the Electric Flow Option

(Nervous System, for example)

This is the flow existed in Human and other types of mammals, principally owing to **Response Time**. *Superfast Response Time* had been a continual need of "Survival of the Fittest". This could be one of the greatest reasons why Electric Signal was used into the flow system and neurons are the carriers of the signals. The potential difference sensed by the Central Nervous System instantly engrosses and sends back the signal through the best accessible mode of electricity.

The neurons are even so designed to transmit signal through dendrites and synaptic joints to avoid any signal mismatch and delayed responses.

The Time of Travel or Response Time:

Electric Signal Flow < Gaseous Flow < Liquid Flow < Solid Flow

Therefore, when we look at the above ascending order, it is obviously noticeable that Electric Signals take much lesser time for travel, and that is why the option was selected by the organism's instinct over a period. The central part of the creature sends electrons via neurons for such movement of signal carrying information so as to protect itself from external threats and any kind of detections, which must be the essence of their Survival or Liveability. The nervous system, Spine and brain essentially operate through electrical flow, and that is the reason the electrolytic balance is crucial and inevitably demanding in an organism and its intricate cellular level. The electricity generation is captive for internal utilisation by the body for responsiveness, Life Processes, Internal regulations and wholesome control mechanism.

The electric flow within an organism ensures ultrafast responsive mechanism and overall control of the body, which flows through subtle potential difference, thus not indulging in any wear and tear like a Heart or Moving Organs. The highest level of prompt action thus requires flawless, minimal wear and tear and less number of tangible joints. All those qualities are possible through the Electric Flow and Neurons. Owing to the fulfilment of functional intricacy and time-bound delivery demand, eventually, the neuron transformed to a desired functional shape.

Fourth, The Gaseous Flow Option

This is a necessary go option for any organism to be fully connected with nature by the incessant exchange of Survival Gases O_2 and CO_2. The lungs or gills or any similar anatomical structures have evolved over time to capture through a transitory exchange terminal by ensuring adequacy in surface area (in and out function) To absorb any gas from the environment, Organisms or Beings have no option to remain than to tap available gases from Air through gaseous mode exclusively actually only one. Thus, gaseous flows equally crucially connect between organism and nature.

The gaseous flow requires a suction mechanism and higher absorption/terminal area where an intricate mechanism pulls and stores as much oxygen or required gas, as the case may vary with respect to different existence under planet synchronal exiting/expelling as much carbon or oxygen from the body. The gaseous flow is an exclusive option for an organism and an obvious connection between environment and organism for its liveability. The gaseous formation and remaining on Earth are the greatness of Earth's atmosphere, and there is involuntary support from the Green planet and Ozone layer as well.

Summary of The Flow

All the four vital flow options were path-breaking in the process of Evolution of an Organism, especially for a species type of Human Being, which took million years to be established or remain to be organised over a long stint. There might have been billion hit and miss during exploration of such process, eventually for Ultimate Survival option; things could have shaped up in the best way or at least current best way, as evolution is a continual process in the planet. In addition to its flow options, an organism might have selected and formed its overall anatomical/structural parts from the same planetary element with the aid of its thinking and reacting mechanism. The flow within the body is certainly not magic rather a selective option chosen eventually in a hit–miss and audacious way by the body, which has a scrupulous transformation right from simplicity to intricacy in its anatomical pattern owing to survival in the purview of protectiveness from external threats and growth contention which is inherited.

WHAT

Sometimes the curiosity is intensified to the Hammering level where an explorer is absolutely unstoppable and can exert a potential of writhing a giant rock for demystifying the facts from the core... Here comes the "WHAT", which may profoundly be referred to "Decoding the Mystery": *The Reconnoitring Journey* from the surface to core, which may invariably vary from having Astounding, Exhilarating, Intriguing, Anguishing, Hilarious (if searching for insanity) and Knee-jerking (if exploring unusual matters) experiences.

The explorer and process may be broader and numerous in existence, but the milestone: *The End-Point is a dot* "...the process is usually witnessing contraction of the events, Space and Time dimension to reach a beginning point with respect to the concerned event".

What Happens After Lfe?

This is the most intriguing question of humanity and *Homo sapiens*, who have developed an ability to think by their evolution. As stated in previous chapters, I have elucidated the purpose of Life Creation. However, the exploration elements of culminating, in parts or whole, which pertain to "after and before life" indeed were never conceptualised and presented in this book so far. I intended to present sensible and unproved contemplations to readers for their appropriate interpretation respectively in desired formats and perspectives.

"The afterlife is an immaculate combination of Energy reversal, Over the air Memories and Pseudo-existence of left out anatomical possessions' and Inclination of Paranormal thoughts by affected physical brains".

With this note, let The God Point start digging from the surface to the core: The Reconnoitring Journey of What.

The God Point is put into action for closer surveillance to unravel mammoth subdued facts and mysteries of pre- to post-life through hypothesised concepts. Let us try to visualise the concepts of such content (despite being hypothesised) which often perplexed and covered with high intricacy or insanity, which might depend on their true factual evidence. This in total would be called a daedal effort (even if may end up without any appropriate coherent inference) when someone attempted for any kind of experiment, research or convoluted imaginations in context of life existence before and after…

The witty part of such discussion or interpretation, whichever reader may feel so, is to apply full assertiveness and thoughts synchronisation with The God Point, to absorb more from most. *The life is an instant stimulus, which was generated eventually owing to energy transcending and transducing into an organism*, the statement in italics is an extract from the previous chapter and elucidates readers about the role of energy before life. However, there was no role of Energy after life, as it seems from the learnings "Energy Transition episode". This is often true as The entity after death is (fully or partially) destroyed by Organic Transition of Planetary bodies. The Entity is, thus, lost eternally into space. In fact, there is no Entity at all. This clarifies merely nonexistence of afterlife concepts.

Is there absolutely zero role of Energy after life? If it is so, how can we justify or interpret: *People's long-standing beliefs, paranormal cases, reincarnation evidence and similar studies or event recurrence.* To understand this, let us take an example of a battery-operated toy. The toy will continue to operate unless the battery fails to retain its charge and subsequently the potential is exhausted. The entity here is a toy, which is existing in this planet and entity may pose itself with life or without life. Basically, it is a binary type of *"on and off mode"*. The Life Energy Transition of such toys is reversible subject to availability of external factors support. In contrast, Living Organism Case is totally irreversible as the energy *out is out. This is a cent per cent accepted concept globally* unless few rarest cases popped up such as recovering from big heart stroke or near-death experiences, where a pacemaker or emulating external aid can help reversibility of life.

The afterlife energy transition process is irreversible, and we can say, "lost is lost". However, afterlife energy may be reversible by an external aid performing re-establishment process immaculately with nanoseconds accuracy. Thus, such restoration does not need any supernatural interventions. The afterlife concepts often perplexed with assumptions, which eventually turned into stronger beliefs, which were interjected into a Human mind at a deeper level. The Entity, especially Human, ponders whenever it/he/she encounters any of such intervention, albeit not a real encounter. A pseudo-encounter is adequate to influence Human mind abysmally, thrusting apart from reality, by making them compellent to be in a pro-clairvoyant state. The brain has a very strong capacity to retain every piece of information and this *"strength"* indeed brings more intricacy in one's life. However, retrieval of memory is not comfortable and sometimes such *"weakness"* helps in healing from mental trauma. This clarifies, the Human brain has an iceberg role that scarcely one could realise instantaneously. Unfortunately, no one can access GOD POINT to see a clear fact happened in the past. Thus, suspense remains intact eternally for the majority of paranormal or supernatural encounter cases. The thraldoms in Human life bind people and affect them keenly; therefore, Human prefers to be in pseudo-world involuntarily. It is the paradoxical nature of the intelligent Human brain to have two different shade contraries to each other. This was obviously developed as an inseparable and embodied part of Human brain as a result of Evolution and Re-evolution Process. The Time and Space as explicated in the first chapter can never repeat because of ingenuity.

The Afterlife Hypothesis: God Point View

Life is purely Energy Transition and an organism is a true conductor of energy. As long as the organism has the ability to receive and partly generate, converse, retain and transfer Energy formats, there is life within it.

The energy gradually stops conduction from the organism or Subjective Entity by making it immobile eternally. The Physical Organic Possession of an Organism is destroyed fully or partially by the external microbes and other assorted factors. Therefore, the immobile organism evades its existence, thereby paving the way for decomposition and fossil remain. Finally, the leading conclusion is Entity must have Holding and Transducing and conducting energy.

Let us get into scientist insights of this.

The state of afterlife is medically an eventual death of all body cells. However, in religious and Cultural and Spiritual context: The state of afterlife has another transition sequence of Energy. Let us assume both are true for further analysis or devising without taking into account of factual information or evidence.

First Side "Spiritual *analysis*"

The analysis ranges from before life to afterlife. The Most Spiritual Context of life starts with soul, which has a further division of Holy, Supreme, Good, Evil, etc. This is too generic idea extracted from various religious testimonials irrespective of religion and region. Now readers have to interpret The Energy Transition Concept with respect to Spiritual Idea. Let us assume that the Energy is soul and vice versa for the time being. If soul is Energy, this must pose a format albeit not in physical state.

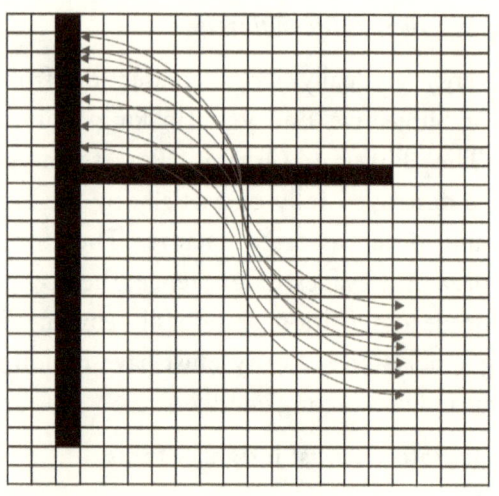

WAVE ENERGY - THE SOUL

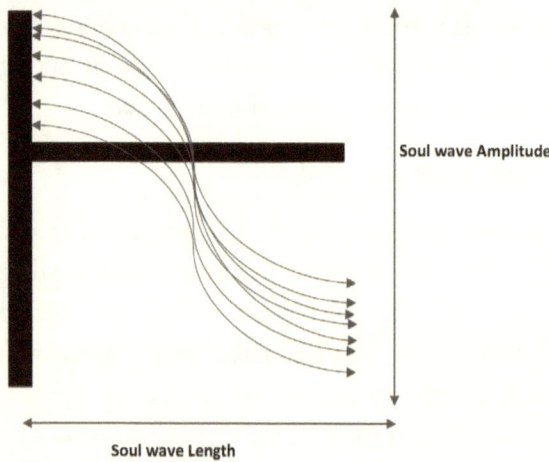

Soul wave Amplitude

Soul wave Length

In this world, we have a visibly predominant and apparently vacant area barring Solids, Liquids and Gases (The States of Matter). Such area is innately not vacant, indeed a medium for all kind of Energy waves travels retains and bombards. The medium is liable for Energy Wave's transition among the Earth, living things and the Universe. The Soul Energy could be any of many such as Light, Sound, Electrical, Magnetic, Heat and Combination or Merged Energy. The soul must carry minute potential and requires a medium to travel. The Energy must be in a wavy format; thus, we name it as a Soul Wave, which has its unique length and frequency.

Let us look at the depicted wave details under 2D "progressively extrapolated to 3D visualisation". The Planet, which has pre-existed enablers, welcomes such wave for interaction. Now looking at the depicted image of Wave within Grid, six kinds of waves are travelling with exclusive features.

Waves neither merge nor interact with each other. This is why it is ascertained that each soul is different and chooses a body during blastocyst (Hypothesis). There is no scientific explanation for Entry time where Soul Energy is embodied for a physical identity. However, The God Point has explained a concept of Energy Transition before life incessantly to a non-living creature and eventually turned it into a living one; let us assume that the incessant flow might be a Soul Energy as well. The Soul Energy Transition starts right from before life with unique features and ends afterlife. Thus, if at all, The Soul Transition occurs as per Spiritual way and ever discovered in future, say 2150 AD or later, this should be demonstrated with impeccable scientific explanation.

The radical aspect of afterlife existence is memory. Yes, the memories which make more believers than sceptic or vice versa, wherever a particular case slants momentarily. The moment here is Space and Time, where/when the event occurrence may be in support of a case or quite opposite to the case. The memories of any creature, while it was living, are stored in brain. Someday or other, when the creature dies it must lose its physical significance. Then why the afterlife memory is retained and if so where is the exact location. Scientifically the situation is incoherent, and it is impossible to retain the memory without physical mass holding. However, Spiritual-side and Paranormal experts have an impactful upholding of such strong beliefs. The concept of reincarnation, evil spirit, etc. still runs the fiction world of stories, imaginary films, few real-life video captures and lastly the scientific measurement of spirit activities. Thus, it cannot be ruled out without subjective evidence gathering and insightful understanding. As we are still in the subchapter of Spiritual-side analysis, let us continue as per the prescribed and stipulated direction by The God Point: the Spiritual-side existence might have been there but unnoticed by science due to inadequacy of the knowledge and technological advancement or all together overall science direction may be flaw or deviance. Let us open up the mind then. To view an afterlife entity from The God Point, the tracing of memory and soul energy should be in cards. That is where the appropriate analogy instigates. How the memory is stored in the brain during survival, taking the case of a Human Being?

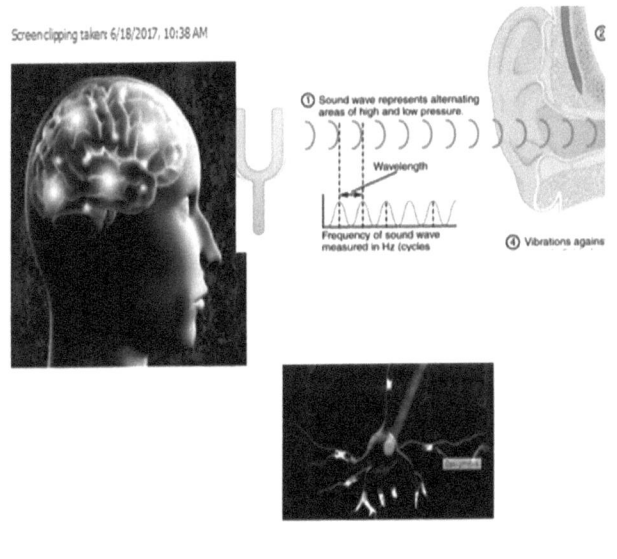

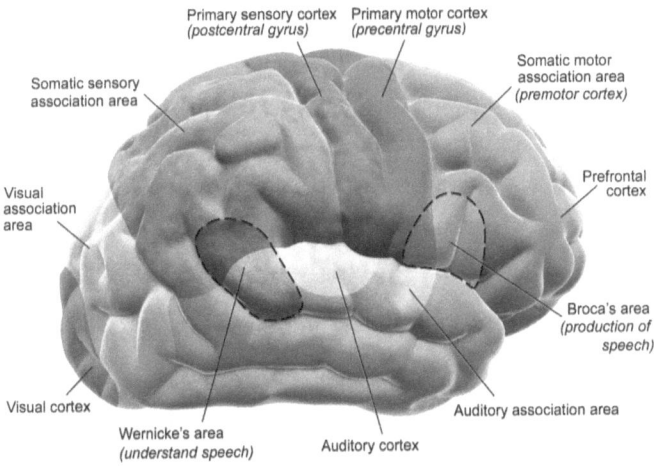

Pragmatically speaking, the Human needs a physical form to store memory. The memory is an event of Time and Space, which is captured with respect to any or combined modes of visual, audio and sensory feelings.

These are basic event constituents, which are indeed converted from sources of external energy. Such external energy is transmitted from the external environment through various types of Energy waves into Human through all sensory organs of the body. The Human decodes it chemically and electrically and stores in the brain. This is a widespread fact that Human brain never loses out any format of stored memory from "within it" unless destroyed or retrieval process is stalled.

The confined memory, of course, is an eternal entity. The memories remain forever through brain cells, albeit, sometimes not being retrievable. The storage of biological format of memory may be analogous to magnetic memory of electronic storage concept, and this concept is generalised and chosen for perspicacious understanding.

The biological format of memories is indeed chemically and electrically decoded for eternal storage physically and this is the most crucial point to be captured by readers, which implies that:

"Human *or any animal can store intangible parameters in tangible formats for almost all biological aspects of life.*"

This is where life starts unfurling and unfolding, thereby encouraging keen penetration of analysis on pertaining subjects. In line with the concept of biological intangibility conversion, we are applying such intricate concept into the Life Energy Transition by means of conversion into a pre-defined Energy Wave of suitable frequency, wavelength and noise. *Thus, the wave can be renamed as*

"Life Wave" or "Wave of Life", which is nothing but the concept of soul in the context of spirituality and religious beliefs. The moment The Life Wave initiates for further conversion in an organism through a Spark is the point of life generation. Hence, life is formed, not by a simple process, by hit–miss way over a million times. The previous chapters clarified the Energy Transition and intent of life formation, so a further explanation of the subject matter is redundancy and can be abated.

In pursuit of reconnoitring afterlife, before life is clarified to a moderate extent. Therefore, the ideology of spiritual context of before life and that of scientific context of before life are *same and coinciding*, while only existence is concerned. However, the absolute reasons and characteristics or Traits of Life Wave cannot be correlated without adequate evidence and astute scientific studies in related matter thereof. The coinciding nature of The Life Wave event is impeccably evident owing to clear existence of humankind today. The approach of clarification and records may be different, but life and planet are the same for all.

The life wave thus instigates, multiplies and allows the organisms own ability to create, store and channelise energy. The life wave stays with as long as the physical possession can retain it; however, the retention is never eternal. The life wave transcends back to cosmos; however, the transition process does not seem to be very astute and coherent with respect to an established scientific justification and illustration process. This concept has to have astute understanding by analogy waves and their memory storage abilities and overall synchronisations.

The God Point

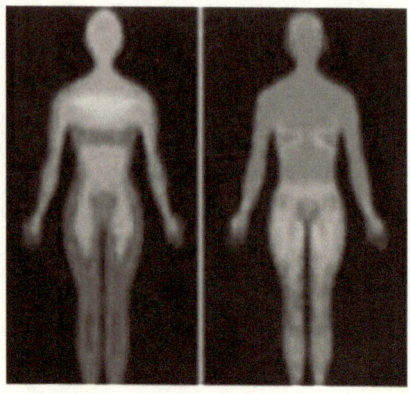

The timing of astral disembodiment in which the spirit leaves the body has been captured by Russian scientist Konstantin Korotkov, who photographed a person at the moment of his death with a bioelectrographic camera.

The image taken using the gas discharge visualization method, an advanced technique of Kirlian photography shows in blue the life force of the person leaving the body gradually.

The Concept of Soul Signal

The signal is an indication of Life Wave presence or movement through the transition of energy flow. Likewise, a cell phone signal operation by keeping air as a medium wherein there is no physical visibility of entities is absolutely not obligatory. The events are memorised in a biological format as long as they are contained within a physical body. The same, however, transmits through an Energy Wave to the external environment as the need of expression comes by the body or brain. For example, when someone wants to share his experience to external spectrum, he/she needs to retrieve and convert the biological stored form of memory to sound or any emulating format to express. Thus, the memory is converted and transmitted through an Energy Wave.

Therefore, the transmitted memorised experience can be stored and retrieved even when the person's biological existence is no more. This elucidates that the physical format is needed sporadically or never to store once the energy which has already transmitted into the air or to the surrounding. The memory can be utilised and manoeuvred or else modestly untapped and eventually drained into the surrounding forever, anything that happens all under contingency by the surrounding and random factors. On a similar note, we can extrapolate such concept to overall key energy of life, "The Life Wave" which can be obviously amassed in a wavy format before and afterlife transition. Eventually, it has a significant virtue of frequency, length and coded form of basic memory to instigate life process once it has to flow into a basic cell ready for division. Hence, the soul "The Life Wave" might possess some memory and transmit to some entity of Earth. Thus, memory is retained for a while in space and finally decoded into an entity, which has the physical significance of memory storage. Technically, this should be applicable to all life formats; however, this is a purely hypothetical concept as of now to justify spiritual version acceptance.

The Direction of Signal

That is obviously unpredictable, but has a pull from objects for necessary, through random interventions. The conjecture about the direction of the soul signal may be following its physical virtues like frequency, wavelength and coded formats to find a direction and culmination. Such a perspective form is solely hypothetical and needs vigorous study in scientific approach instead of being too much conceptual.

Nevertheless, such hypothesis somehow makes persuading idea of spirituality on Soul and Life Transition before to after. The stored memory the "Life Wave" Carries is still indecision, which cannot be predicted so sooner but if such concepts ever existed, it must be required to demonstrate them scientifically. So far Spiritual and Religious beliefs are persisting, this may emulate and justify to a smaller extent. However, such queries of life after and before still maintain their intactness of being mysterious topics of the Planet.

Summary of First Side "Spiritual *Analysis*"

The idea of spirituality and pre- and post-existence can never be ruled out, concurrently, that cannot be accepted as well without definite evidence and scientific proofs. However, the concept has the potential to prove itself over a massive time scale count or period in future and accordingly I have expressed my conjecture to have further intuitions. That may be an open-ended substantive decision of readers to interpret or disbelieve, whichever they may feel so.

Second Side "Scientific *Analysis* of Pre- and Post-Life"

We really do not need to have an elaborative, extensive and contemplated approach for scientific analysis for insightful understanding in reference to pre- and post-life aspects. This is because birth is the beginning and death is an end. However, there is a catch of birth and death. The birth mentioned above is factually right from a zygote or an individual identified cell after mating of parental cells through a biological urging process. Upon extending in a similar line, the death is scientifically death of all the cells and the body has the inability to store any energy and is open for decomposition by external cell eaters. The conjecture on life after death is basically energy retaining inability and eternal irreversible Energy Transition between the being and surrounding. This is the Energy Transition right from the beginning to the end and eventually back to the Universe. The state of death is an inability to intake, produce and retain energy by the biological body. Thus, the state has limited time to restore its state to make reversible otherwise almost majority of cases fail to restore, and the death is the outcome of the irreversible process and is the channel for a way out eternally.

Summary of Second Side "Scientific *Analysis*"

What happens after life: The unwinding and unfurling secrecy is very difficult and can never be answered unless someone encounters and ensures a reversible transition back to life; it can be termed by The God Point as,

"The life after life biologically not spiritually."

This does not mean the secrecy will never be shattered by emulating evidence. However, the contingency even without a specific encounter is still apparently doable with the strong support of technology, which is the essence to unfurl such an intricate mechanism of life. However, to end with decisive facts, I put my conjectures as below for readers' interpretation.

- The Energy Transition to the Universe is certain after life.
- The reincarnation concept is ruled out due to physical destruction of the memory of previous life.
- The time travel is ruled out due to the inability for apple-to-apple physical transition biologically.
- The cells are last to die and skeletons to remain forever. The total organic possession of an organism is dethroned back to the planet.
- The planet is the true parent of any organism.
- There can be energy retrieval and exhibition of memories but that is not going back to bring back The Life Wave, which indeed never existed scientifically.
- *In one word, there is no life after life.*

The sovereign of all intangible parameters of the planet is "Life Energy itself" which is never palpable and unconditionally divine. The existence of afterlife spiritually is the exhibition of life even after the biological life, thus there comes the law of conservation of energy. The energy is eternal and indispensable. Sometimes the pre-existence conjecture can be more valid: There is no need of any Creator when energy is pre-existed in this Universe. The "State of Time" has arrived now at this point @Universe coordinates for going into "Summary and Conclusion" ...

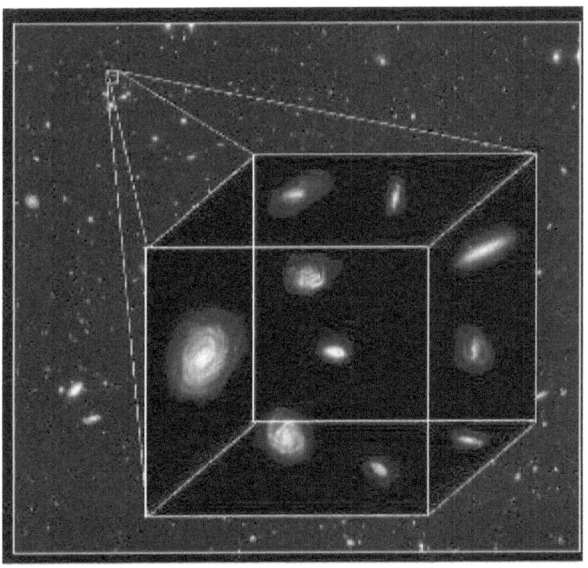

"The pre-existence indeed is the basic state of the Universe, which has to be in the state of a filler of a large giant void without any alternative choice. The Universe what we understand today is a subset of a Large Void which is eternal, infinite and partially invigorated with a scope of perpetual expansion."

CONCLUSION

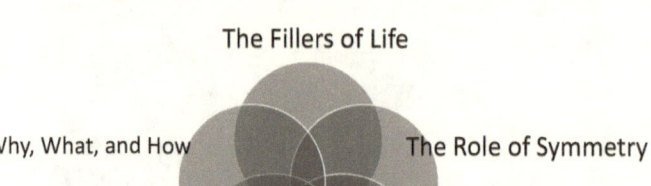

The conclusions are commas but not eternal, full stops in a sentence. Likewise, this is not the penultimate resolution of everything, but can be something of everything or everything of something. All those depend on the cyclic pattern and event chronologies for futuristic aspects of life. However, for The God Point 1, here we phrase out conclusion. The multi-perspective approach and view from The God Point have laid down some important life aspects in conclusions. The propensity of sticking to the past always affected Humans' present and of course future actions. This shortcoming is episodic forever by strong determination and consenting the world as it is. The conclusions will devise the "dos and don'ts" of life from the perspective of God Coordinates or God Point 1. The God Point 2 will be my next book (a sequel) which will exclusively soak contemplations about God supported with a fictional concept.

Now, let us rewind the chapters dealt so far, where we picked five major concepts.

- ✦ The Fillers of Life and Entropy
- ✦ The Role of Symmetry
- ✦ The AGE – Anti-Gravity Effect
- ✦ The Flow and Human Design
- ✦ The Mysterious Life
  - Why, What and How *The God (in God Point 2)*
  - *The Conclusion of God Point 1*

The excerpts from conclusion are something, which have to be taken sensibly and possibly put into practice by the readers. However, those are left to individual choices and can be apparently insignificant unless understanding insights and derivatives thereof. So far, the book has covered all basic instincts detailing and intuitive ideas pertaining to life. By the previous chapters, we can briefly jot down the role of fillers in life, why fillers should not be taken so seriously, how the Symmetry and Balance affect individuals and their connection with natural aspects like gravity and peripheral environs of planet, how the planet is important, how energy transcends from the Universe to plants and mammals, purpose of Life Creation and Finally What happens after what and if so, How! All these have devised, "Why Humans should correlate the majority of things with respect to science. Why the time what we spend must be devoted to the true crucial function and Human must come out from pseudo-possessions." Finally, how to make our life meaningful, beautiful and joyful and a smoother exit. In other words, no regrets about life before and after and even in the phase of Transition.

The Conclusion is an Inevitable Beginning

The perspective of well-beginning definitely requires a meaningful conclusion of the previous episode. That is where the life transition points have an appropriate précis and efficacy in synaptic trades, albeit not fiscal. The more efficient the exchange or transfer, the best the beginning. The life transition is the most thought-provoking and that is where the term change comes. The *Changes* can be either way and often described as a great or miserable transformation, that all are a contingency of consequences and dependent on the state of mind of besieged people. Let us start the conclusion with Change. The change is any incremental notable difference even a small activity and dimensional positioning displacement with respect to time; this may even be able to be osmosis of cell or neuron synopsis. This is traceable and sometimes seldom traceable with respect to available state of technologies across the globe. Human reacts to change voluntarily and involuntarily.

Conclusion: Through the God Point

The Vital and Vigour five aspects, as mentioned below, are the conclusion and advise by The God Point for readers to secure their life better than ever.

- The "2L"
  - Pseudo-possessions
  - Memory cross-section
  - The Emotions and Predictive thinking
  - The Time
- The "Change"
  - How to interpret it
  - Thinking "Out of the box"
- The "End"
  - Why are we not Happy? (In God Point 2)
  - The end is an End
  - The Quote

The 2L

2L—*Live and Leave*. We exist between two terminuses and adhere to a stint of time, which is quantitative as age and qualitative as Happiness. This planet is a dwelling support where Humans Live and Leave. The Concept of 2L is a tale of the detrimental distance between two extremities, which shortens as it advances and converges to a dot. The dot is indeed a "Merging Dot" of two endpoints and a definite inadvertent milestone of Life. Amidst space expansion and time enervation, Human encircles a birth and death cycle. In between **2L**, we have many possessions including organic, the body of a Human. The Concept of 2L may be an age-old idea and offers no new understanding on people who have adequate spiritual cognizance. However, such concept applications are always refreshing while implementing with contemporary lifestyle. The measurement parameters of "Live and Leave" can be equated to "Age" quantitatively" and to "Happiness" qualitatively and to "Merging Dot" Ultimately. The Ultimacy is an eternal end, as right as rain and bolt like blue… The dot eventually sublimated to the Universe, through an Energy reversal process and in pursuant of maintaining never settling celestial equilibrium. As the evolution process is iterative, Human life cycle is redundant; we must *Live* with possessions, Imbricated Memories, Emotion and its Contagion, State of mind and A Certain stint of Time. In a similar manner, we must *Leave* all while death is underway.

These are the inevitable crucial aspects of life:

THE 2L CONCEPT ILLUSTRATION

LIVE--------------------→.←--------------------LEAVE

LIFESPAN-LINE

LIFE SPAN = Time as Quantitative Measure

HAPPINESS = Qualitative Measure

THE END = Merging Dot as Ultimate Measure

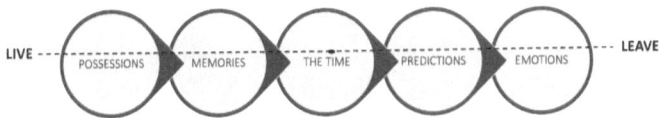

The explicit ideas about those five aspects are:
- ✦ Pseudo-Possessions
- ✦ Memory cross-section/ Memory card model
- ✦ The Emotions – Happy, Sad and Joy of Gifting
- ✦ Predictive Thinking
- ✦ The Time

Pseudo-Possessions

"The possession should not entangle us to a stage of perplexity propending destructive ideas and uncontrolled intended deliveries."

The filler stuff between 2L is possession, in fact, pseudo (false) possession. The pseudo is indeed the entitlement of all Earthly Possessions. Therefore, an entity can be replaced as Ps, i.e., "The Pseudo" instead of salutations as Mr. or Ms.

The God Point

Hence, the identity of anyone can be rephrased as Ps. "Name".

For example, Ps. Soji Queen or Ps. Sheela or Ps. Whatever...

The existence is an infrangible state of time but not eternal owing to temporary possessions, we accrue right from our zygotic inception and over a long stint of time. The State of time is not that long if compared with the Time at ISS (International Space Shuttle) or any large planet. When Space and Time are variable, there is no point in thinking of stability in one's life, the inevitable change that eventually befalls. The transfer of ownership on the planet is almost a pseudo where no entity firmly upholds their positions and belongings eternally. The life essentially needs possession of a stint, quantitative matter of planet such as organic and inorganic, biodegradable and non-biodegradable entities, which eventually are reversible and transferable. In contrast, possession is also inherent and a must-have aspect without which our identity is null technically. What we must keep in mind is "The possession should not entangle us to into perplexity, which has a propensity towards destructive ideas and uncontrolled intended deliveries". The possession inflates basic selfishness to imbricated selfishness, which eventually slopes towards coupling with inhuman behaviour and hence, such inflation can be devastating sometimes. (Basic Selfishness is an Essential Attribute.)

The takeaway from the above paragraph:

"The essential trait "possession" must be kept in check to a degree demarcated by a finite distinct line between Obsession and possession."

The God Point advises discerning the line at the earliest for absolute peace in life. The possession eventually penetrates profoundly in mind and instigates relativity in the level of possessions and accruals, which may be skill sets or any virtues except inherency. This phenomenon is comparison in simple terms. The comparison is introduced out of possession. The comparison is devastating when it turns compulsive and ultra-reactive. The pseudo-possession leads Human into two ultra-reactive behaviours such as being competitive and compulsive.

The Comparing nature within Humans cast to emulate possession level, which may be due to various socio-cultural aspects; sometimes even a young child may be too possessive about zero-valued or quasi-garbage items. However, the right to compare in a sportive way is always essential owing to good and bad differentiation, proposing improvements, Strengthening the individual capabilities, etc.

There is a certain level of maturity to be put in action before engaging into scenarios and timely discerning will help everyone for stabilising their life if unstable and retaining peace, if stable.

The materialistic aspects must be denounced from Human mind as those overthrow your peaceful life by infusing obsessive possession and unjust comparison incrementally. Even justified comparison is vulnerable to reactive behaviour. Thus, stay away from possession, which is absolutely of no need and like greed. Society will never be happy with you, even after you die. If this or that comes as a prefix eternally in everyone's life, make sure to devote life only as per your requirement and gratification. The Happiness Chain must be cut short and standstill with your own feelings and be yourself, which may not lead to supremacy or legacy but an immaculate existence of yours in this planet.

The Memories – Memory cross-section

The 2L – - "Live with memories of past and present to create present as a memory of future; Entity should Leave past Memories and go forward until entity Leaves Life."

As stated in the "How" chapter, the memory area of the brain is Hippocampus. The brain never erases memory unless destroyed by some means. Fundamentally, memories are stored as microscopic chemical changes at the connection points between neurons in the brain. Neuro Scientists say that "The brain contains 100 billion neurons, each of which connects to up to 10,000 (typically) other neurons. There may be 100 trillion connection points or "synapses" in the Human brain. As information passes through networks of the brain, the activity of the neurons causes the connection points to become a disposition in their state of response. The strengthening and weakening of the synapses are how the brain stores information. The hippocampus is specialised on coding and structuring memories, particularly autobiographical and episodic memories (memories about people, places and events). However, some scientists believe that memories are only held in the hippocampus temporarily and are later re-coded and dispersed throughout the rest of the brain using a process called "memory consolidation", which may happen during a subconscious state. The precise way that long-term memories are structured and represented across billions of synapses is the subject of intense ongoing research and remains one of the great mysteries of neuroscience. Location of the hippocampus specialised for episodic memories... Statement from various neuro articles.

The Biological Memory is the most intriguing aspect of medical science, and the study of memory is still at its nascent stage.

Now, we are going to understand How 2L is related to brain and Memory and Possession. To devise more on it, let us put forward insight interpretations. On prima facie, Human needs certain form of physical existence, which is organic possession, which is obviously not only reversible physically to the planet—the Earth but also reversible intangibility to Universe through perpetual and Inevitable Energy Transition. The memories remain forever, even though the planet reverses Life Energy and releases the body from confinement. However, memories must need any physical format for storage and retrieval as well.

The astute brain keeps everything intact inside the multi-layer functional and structural mass. The functionality has a close relation with structural evolution, this is explicated in previous chapters. The way we script our next path depends on the past experience, present state and pseudo-possessions; thus memory plays a vital role in all state of time, events. The memory and intelligence both are synchronic to each other. People often rely on memory as a chip-based biological system in the brain, but Human is different package altogether which has infinite RAM, Infinite Processor, Infinite Storage and Infinite Creativity acting under same space spherically and harmoniously to follow a specified rhythm like an advanced feedback control system with an incessant command, learning and implementation processes. We need to comprehend our brain and memory in a simple way without getting into intricate ideas, which can be through biological dissection and ideas proposed by many scientists, neurologists and doctors in the past... The point of contact between memories and emotions is Human intelligence, which decides to function with many probabilities under different scenarios.

To interpret this, a memory chamber model is depicted in figures F1 and F2 through a cross-section of the brain. *The supremacy of Homo sapiens today is owing to humungous storage, lightening retrieval and infinite processing ability in an organic, delicate and charged mass, the brain.* However, this is a part of evolution and took five billion years+ for shaping up to a desirable state of time.

The memory model suggests two kinds of memories:-

1. Simple but imbricated
2. Intricate and Imbricated memories

The intelligence studies memories from – Infinity to present and Present to +Infinity even for minor to major issues of life, this is biological aspects of brain, which is pragmatic in its way... The range of infinity (+/-) indicates the state of time right from inception to present and future based on predictions and creativity thereof.

The evolution surpassingly relies on brain function and genetic ability thereof. Thus, superhumans are not commonly inherited as offsprings even today despite incredible technologies in place. The Contemporary Humans still do not exploit their brain to fullest and pass the virtues to genes to create a real superhuman.

(Contd.)

The Brain Cross-Section and its illustration through Memory Model: ↘

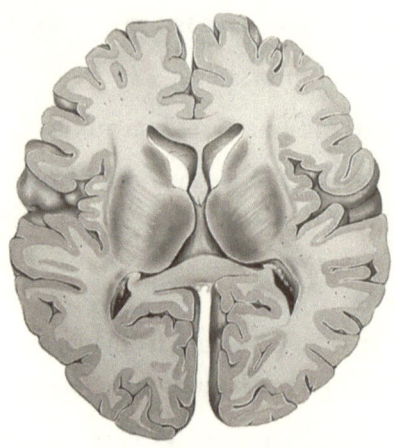

Fig F1

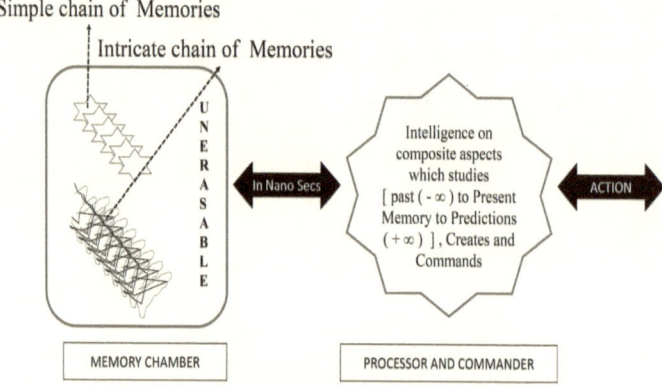

Fig F2

The Memory Chamber and intelligence are coexisting and acting synchronously with actions and reactions in Nano Sec intervals; thus Human is a developed species today... The depicted illustration consists of abysmal storage and processing masses of Human brain, it is ostensibly easy to comprehend though remains unfathomable, if we kept on following the series to reconnoitre source details. However, we should stop at the appropriate state of time and events for needless Herculean task or we may need quantum support to unravel source. The source is not that crucial when compared to the basic level interpretation of radical process followed, which is adequate for the elucidation of such a concept. While the mystery remains always a mystery, unsnarling, to certain steps, both chains simple and intricacy one by one and interchanging them will form another type of memory link—A Hybrid link... The intelligence will have no effect as it can process infinite memories; however, the behaviour pattern is susceptible to changes for the Being or Entity. The Hybrid illustration is as follows... Though storage is important, retrieval is a supercritical aspect for behavioural variations and perfections. Therefore, some people calculate fast mentally, and some are not, some can memorise promptly, and some are not. All such virtues depend on retrieval efficacy rather than storage capability. An animal may still have storage, but the retrieval and intelligence are seldom high as compared to the Human brain, which overthrows them from evolution race... The RETRIEVAL sequence can be altered hypothetically but this seems unalterable at real and pragmatic conditions.at least, as on today. The memories alteration hypothesis is stated below for readers' understanding.

The God Point

A, B, C – Simple Memories chain

X, Y, Z – Complicated Chain of memories

For concept illustration and subsequent clarification, let the terminology AX, BY, CZ be Hybrid Chain of Memories. Now let us decipher letters to actuals with an example. A person's observations, while he was travelling and started noticing almost everything, comes through the train... In 2019 ...

A – Tree X- Taller Than Tree in 1989 at Jerusalem

B – Green Y- scenic than my town landscapes

C – Tall Z- Weather suits this plant

Hybrid: The Hybrid could be various combinations of A, B, C and X, Y and Z in general, however, the selected type of Hybrid memory is stated below for readers' better understanding.

AX – Tree Taller Than Tree in 1989 at Jerusalem

BX – Green and Overall scenic than our town landscapes

CX- Tall and Weather suits these plants

The memories retrieval is usually involuntary and voluntary is by choice of the individual. The combination of voluntary and involuntary is a typical example of Hybrid. There is further elucidation to memory model and Hybrid Idea with a simple example. The illustration as depicted in the next page is based on the "pages of any book". The book comprises content filled with bearing pages with identity. The accessibility of any page at any moment is possible by any individual; however, The Opaqueness of the content is inevitable for most of the year's old cases or sometimes very recent case. The legibility of pages depends on how best the individual interpreted during observation and the quality of Memory Retrieval Process.

Simple and Sequential Memories

Intricate Memories Usually Random

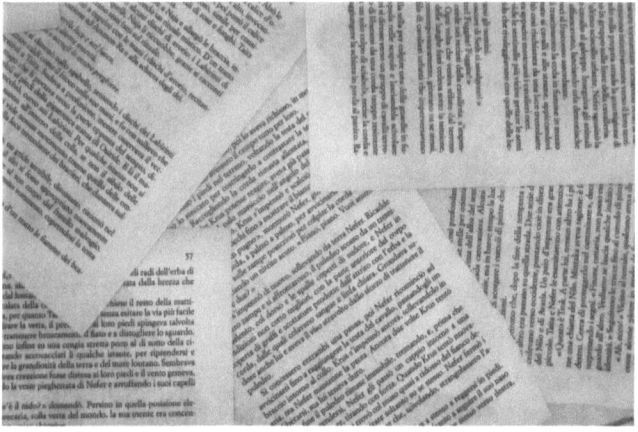

The Hybrid is an astute combination of both above, a similar analogy can be correlated to a random collection of numerous books with their sequential page numbers.

The Hybrid memory occurrence is very common in Human and that demands them for posing incremental intelligence. This is also obviously a specific cause of the "Human Evolution with a purpose." Now going back for memory Alterations or Transfiguration or Mutations.

In fact, the exemplifying person noticed and remembered everything right from simple to intricate. However, for experimentation purpose, AX, BY and CZ are the final memories with accurate memory storage and retrieval sequence without any changes.

1. AX- **Tree Taller** Than Tree in 1989 at Jerusalem

    Behaviour pattern – Wow, Surprise, after a long time

    ↓

2. BY- **Green and Overall** scenic than our town landscapes

    Behaviour pattern – Sense of Beauty, A feel-good factor

    ↓

3. CZ- **Tall and** Weather suits this plant

Behaviour/s – Logic and reasoning, copying the interpretation. The above arrows are directed to the next events, which is a sequential one and systematic arrangement of the memories in the brain.

Thus, the retrieval should be easier to moderate, depending on the event significance and interpretation or sense of urgency by the subject. In contrast, The Hybrid memories trivial subject to individual capability and surrounding as well. The alteration can induce further complications in behaviour, albeit, not an astonishing fact.

*This is owing to a correlation of the quote: "**The divergence of smaller deviation can humongous built a mammoth as the Big Bang built the Universe. This is absolutely natural phenomena existing within everyone (a Convergence induced divergence)**".*

Now by Applying Memory Alteration Hypothesis:- When the Sequence will be altered as BY, CZ and AX, the result is as follows:-

1. BY- **Green and Overall** scenic than our town landscapes

   Behaviour pattern – Question, Rethinking and retrieving attempt.

   Introspection Changes: *What is this green and no feel-good and beauty sense?*

   ↓

2. CZ- **Tall** and Weather suits this plant

   Behavioural Changes: – *Why tall is an issue here and do I really know about the relevance of this plant weather suitability?*

   ↓

3. AX- **Tree Taller** Than Tree in 1989 at Jerusalem

Changes – *No wow factor* and *the entity perplexed with "Why am I getting this tree now in my mind?" and "Am I okay?"*

"Unanticipated behaviour can be partly noted."

Therefore, the Alteration process can totally change the behaviour of a person for the same instance within the self and with respect to the surrounding crowd. Thus people encountered memory leakages or may have such similar dysfunctions can never behave optimally and often be terribly excited for a penny objective. The final line to say is "The Retrieval and its sequence are crucial than storage".

How the above memory model concept is related to 2L

The life is filled with a mix of all emotions, therefore a person needs to paraphrase and prioritise. There the retrieval and behavioural sequence matter most. To ensure maximum concentration during any activity, The Entity must optimise Memory Retrieval Process and should have to restrict the complexity of the thoughts up to a job area or any desirable skill set for onsite application, comparison, etc.

Are we all doing this? Technically few, other common people often have micro to major distractions forever. Distractions are certain in occurrence due to impeccably capable and dynamic brain. The memories are often linked to possessions, and few make us happy and few are certainly driven into a state of misery. When possessions are not eternal, how can they affect your life by retaining memory fallacy of memories encumbering into deep sorrow or acute perplexity state, thereby influencing Human behaviour misrepresentation? If certain memories Type 1 make an entity, excited and certain memories inspire him/her. Certain memories Type 2 take into deep sorrow or deep anger or sometimes do nothing. Can we have deleted Type 2? Obviously not ... Can we alter the retrieval of Type 2, Possible Yes! and here is the Key to Happiness. It is better to go near the object to catch hold instead of stretching hands elastically, which may induce strain or any kind of uncertainty. When we are aware of the consequences, why do we attempt? This is because of the memory retrieval experience of any successful attempt of the past. Take this example to Life and Death events chronology and the same retrieval alteration formula applied to keep someone immaculately happy despite a deep sorrow state in reality. Thus, people need to optimise memory retrieval, behaviour with respect to past experiences and intelligence on future predictions. This is the take away from subchapter memory model and alteration concepts, which are not necessarily scientific, but they may be useful in real-life application.

"Don't hold memories too obsessively and go with the present. Eventually, the present is the past of future and present will still have regrets (despite you put all efforts for doing best) to improve the future actions, understanding purview and Maturity Exhibition. As time passes, The Incremental experiences, cognizance and intelligence drive us to success steps and involuntarily, we must leave all the success attributes before merging to a dot."

<u>Yes, the Dot is a</u> ". " (Read dot as a full stop)

Relativity in Happiness

There should be optimally a meter, which can measure Happiness. This appears as hypothetical as on today but not an impossible concept. Let us start devising what is happiness and How to increase Happiness Quotient through Happiness Meter. There must be certain parameters, which make a person happy in all respects. We are not going to evaluate so many parameters and put emulating justification for each; however, we are devising though blemishing parameters to HQ. The complement of happiness should be the pragmatic attributes, which do not allow us to remain confined in the State of Happiness. The five important complements as per my conjecture are.

1. The relativity in happiness
2. The pseudo-comparison
3. The fear of uncertainty in future
4. The emotions, selfishness and inappropriate action/reactions
5. The memory and its retrieval

None of the above five allows us to remain happy. Those five complements take to a state of despondency and Confusion. The perplexity state is "what we know and how we act and react". This is coupled with a contradiction of "what we don't know and how we act and react, eventually, when we know they don't know"… In this subchapter, I will explicate about the relativity of Happiness only and rest (2 to 4) is kept for readers' own interpretation.

The memory and its retrieval are already briefed in the Memory subchapter.

1. The relativity of Happiness:

The relative happiness index is what we contribute during a mass survey and data compilation is subjective along with some empirical presumptions during the overall ranking phase.

The Einstein Formula of Energy and mass relativity is sporadic.

$E = MC^2$, Einstein's theory of relativity...

Where E = Energy

M = Mass of the body

C = Speed of light

The above equation has a proximal relation to our happiness and relativity. The equation has explicated about the equality of Mass and Energy at some point of Time and Space under certain conditions. In analogous to the above relative equality, albeit not in a normal scenario, our happiness and randomness also correlate with an organism's mass. This is solely based on my imaginary notions, which does not have any scientific significance. Put in usage for illustration only. My conjecture on relative happiness based on the theory of relativity and the general sense of relative aspects:

1.1) The Relative Happiness:-

The happiness in Human is relative, not absolute. Therefore, Human needs a basic comparison methodology for happiness measurement. However, this is utterly a fallacy and Human is sometimes very happy for no reasons. Thus, Absolute Happiness not only exists in spiritual leaders but also for a common person. But the common man takes realisation time and by the time He/she realises, they are entrapped in other Life Fillers, the fallacy of misery or day-to-day life engagements, though insignificant, as the advancing time never waits in general, except atomic clock under varied conditions. Let us not discuss time travel now.

1.2) The Energy for Libation and Absolute Happiness:-

Truly speaking, We Human eventually need to Liberate and seek for Absolute Happiness. However. The Life Fillers are always being successful impedance in life, and we encircle (often prefer to) between problem to problem.

The radical and common assumption is "problems are part of life". Honestly. I feel problems are not at all part of Life and Life can run without problems. The job will have challenges and maybe puzzling for a while, in the process of deciphering or retrieving, but they are never been encountered with true problems technically. The problems are rare and monumental, if they occur, like natural calamities or supreme level inadvertent and uncontrolled uncertainties. Someday, someone would have thought of liberating and leading a life full of happiness. If at all such things occur, Human needs to spend Energy for attaining such a state of ecstasy, maybe a hypothetical state for understanding. Let the Energy be called as E, the energy for liberation and happiness and is required to attain a state of ecstasy. To formulate such energy, let us rewrite Einstein's equation in different real-life terminology.

E = Energy → The energy required for Happiness, liberation

M = Mass → The mass of the body

C = Speed of light → The rate of enlightenment

Upon summating them and putting in Einstein Order;

$E = MC^2$

The E for Happiness and liberation = Mass of Human X (The Speed of Enlightenment) ^{2}

Note: The Required Liberation Energy thus varies from person to person.

In simple terms, not going with intricate ideas such as enlightenment and its rate etcetera, one can say The Happiness and Liberation Energy equates to The Organisms mass × Speed of Travelling through time; thus rewriting yields as below:

Energy for Liberation = The Organisms Mass × Current Speed^2

However, if the body of an organism is allowed for being travelled at the speed of light, the body attains Happiness and Liberation immediately due to available required energy, is not this an unjustified ridicule when an astronaut even can travel near to that speed? Yes, but

NO, as the concept here is altogether disparate idea. That states, if any organism/Human (Small age group) can remain in a State of Happiness without distortions and travel at a speed of light then the organism is nothing, but a fully liberated entity or energy filled with Happiness and nothingness solely. The time will run extremely slow almost near to zero as compared to Earth. Thus, the organism returns to Earth after spending 30 years of Happiness and Liberated Life without many objectives, the organism or Human can see earthman spent 30 Years of life already with Mixed Emotions, Pain and Opportunities and Finally few entities would have liberated from Earth already. The Time travelled Human thus have the opportunity to taste varied and a variety of options to dwell under earthly life. Technically speaking, The Time travelled Human would like to live for another 60 Years of Average Human life with stern curiosity and without a pinch of hesitation, even if He/she tasted liberation, Happiness and Life Ultima already. Hope readers understood my disparaging conjecture of free Human and Earthly Human. Thus applying Einstein equation in general, life may give some philosophical idea of Relative and Absolute Happiness. However, such a concept never existed but is presented for succinct illustration.

E (In short) = The Organism $\times$ Speed of Light^2, therefore The equation states that

Happiness/Liberation/Nothingness = The Organism

"provided" the organism travels at a speed of light incessantly without interruption and distortions.

The Human will ever remain in the State of Absolute Happiness if He/she travels faster than Time with the speed of light. If He/She remained in the State of Happiness, they will continue to remain if the Time and Space do not step forward or move. However, without a disposition and with the speed of light, they travel faster than time and maintain a status quo. Truly speaking at a state of Non- Existence ideally. The state is often called the State of Equilibrium between existence and pre-existence. That state which indicates the state of pre-existence of Human is a State of Absolute Happiness. However, are we ready to be in such a state, if possible, even for a short stint? May be or Not, the answering depends on the state of mind and knowledge. This further reiterates that, in reality on Earth, The Absolute Happiness Comes from being reserved, engaged in striving for the single objective of survival with perseverance, taking almost appropriate precautions by Predicting future events and thinking ahead of time… The Overall conclusion is *"Human prefers State of Existence not Pre-Existence."*

Once we exist on this planet or any planet, surrounding will be an undetachable part in our life, and surrounding sometimes is hostile and sometimes supportive and sometimes does nothing unless we poke. The Human Existence is absolutely filled by Entropy/ Disturbances (good and bad and mixed often). Existence state is always a mixed state, which comprises all different types of feeling, such as Astonishing, Happy, Sad, etc. Being Existed means balancing life without choice, however. Life Can be directed in a better way by being reserved engaged in striving for single objective of survival with perseverance, taking almost possible appropriate precautions by Predicting future events and thinking ahead of time and above all humanity must include activity in life, which keeps Human always in a state of High Potential and Pride without Selfishness.

Human will never be Happy (Relatively and Absolutely) unless He/she wants to attain IT by leaving Pseudo-Possessions and Spending Time qualitatively.

The Emotions and Predictive Thinking

The emotions are the type of feeling exhibited by the brain to every scenario happening around and within the self as well. The contribution of hormonal fluctuations and dopamine is also certainly a nontrivial component, but the memories are the Basic Trigger of everything. As stated, earlier memory is never an issue, but it is the Retrieval process, which awakens the settled dust or sometimes the concealed chronology. If there is no memory retrieval, there is a mono emotional state of Human, which is a stable state as equivalent to robotic behaviour. As Human wants to induce emotion through AI into robotic behaviour, the Human himself can never free from emotions despite the capability, which ensures building steps to moon or mars...

The predictive thinking is typically a foreword thinking approach with million possible simulations, though it never equates frame to frame, stated by The God Point. If prediction really emulates frame to frame with future reality, supposedly inadvertently, there must be supernatural inflow and we can certitude about their intervention. A hypothetical view from The God Point states that this is a never attainable reality of the existence. However, this conjecture is not a proven reality indeed, either side.

I would like to state more about the happy cycle in God Point-2 as the content can be revealed synchronously to the narration of God Point version 2.0 fiction.

Time: 24 Hours and Pi of Life

Time is a pivotal intricate thing, which possibly cannot be explained in 500 words or maybe more. Before ideating about the Time, let us get into the radical aspects of Time on Earth. It is nothing but A Day. The day, it's about 24 hours, can also be rewritten in formulaic pattern "Pi" × 7 h, 31 min, 7 s.

Time taken for coverage of two extremities of the Earth, either North and South or East and West. If someone is allowed travelling through a core tunnel between extremities, he/she requires only 15 hours approximately for covering a day if he/she was allowed travelling by the speed of the Earth. However, an approximating contrast velocity over speed can be defined. Supposedly, someone asked to attend a pre-scheduled event next day morning at 10 am. The traveller decided to reach to other end and came back to the same point, he/she might take 15 h in ideal run. This exemplifies about saving of time and attaining early success. Thus, this creates irony about

Time Enervations and Biological Enervation, whether they are densely related or pole apart! For intriguing insights, let us dive into a Time Chart of a cell, the basic unit of life. *The diameter of Earth is 12,742 km and Earth is rotating at a speed of 1670 km per hour.*

THE EARTH AND TIME

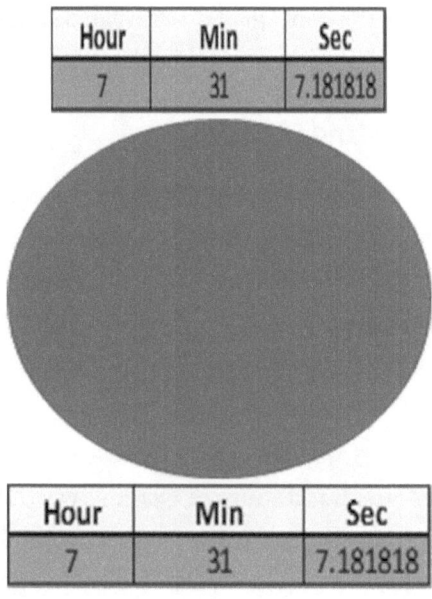

Hour	Min	Sec
7	31	7.181818

Hour	Min	Sec
7	31	7.181818

Time Enervation and Biological Enervation

Space and Time are never ceasing and perpetually expanding in Universe of Universes or Infinite group of celestial bodies. The Human is nowhere in the Universal Map but capable of infinite abilities, explorations and embossing inter-galactic footprints. We, Humans, are concerned about Time but not actually, on Biological Time, which is within us and an impeccable bio schedule.

The Modern and Future Human may travel in Space and Time but can never alter Biological Time through a natural process. In the future, we may be successful in interrupting ageing process but this concept still is ostensible anticipation of reality without side effects and anti-natural dwelling.

To ensure such concepts' realisation, The Basic Unit of Life must be studied extensively. Cellular biology is comparatively crucial to other medical aspects. The cell functionality thus inhabits an impeccable biological schedule within itself and that is the basic Time, we should focus on. The time, which generates future, is unceasing and thus our cell synchronously follows the time of Universe. The time varies with respect to gravity and other celestial parameters, but the Biological Time is near-constant and unvarying unlike The General Time. There is a disparity and observable distinction between Biological Time enervation and Physical Time Enervation.

The objective of the "Time" (subchapter) is to unravel distinctive aspects of two types of time concepts irrespective of event chronology and alteration possibilities. The subchapter draws a perspicacious line between biological and metaphysical times, thereby reiterating us for self-realisation on Time Race (Racing against Time, time travel and Physical Time Enervation aspects).

There is absolutely no need of Effective Time Management except "Pi" rule, which should be followed involuntarily. The three equal parts of a day, 7 h:37 mins, derived by dividing 24 hours with "Pi"= 3,1456. We need to allocate the three equal zones of a day for Work, Sleep, Family and Leisure and Personal Care, which summates to approx. 22.5 Hours. The Balance of the day, the remaining 1.5 hours approx. for Self, Your Self. Indeed, The Last One hour and twenty-seven minutes are a "Time of Choice" for doing nothing or anything, maybe Remaining Calm and controlling our thoughts, kind of Meditation/Yoga and maybe a Time which we should be dedicated exclusively to ourselves.

This time is undefined and can be allocated to those activities, which are not in the tertiary category of Family, Leisure and Personal Care and allocation should not be perplexed with Leisure Time especially. Time Management is an effective skill as long as we feel we are on the right track without any deviation. However, the deviation comes despite all efforts and that is the way of Life. Therefore, Time management should be generic instead of minute-to-minute control, which unheedingly exerts pressure on the Human brain and Affects the copious mood. Readers do not worry about metaphysical time, which is created by Human for Organising Events and Planetary purposes. The incessant watch on time keeps everyone on toes and absolutely induces unneeded stress to the brain. Please follow punctuality rules and time adherence regime to the basic level. The Sun and Moon are true indicators of ancient civilisation and people were hardly concerned about hours, in a day splits; expect Work, Food and Protection. Time has its own significance, but this should not influence Human to a devastating state of Mind and Body. The Time, though linked with events, is independent with Human behaviour and Happiness. While Time never waits for us and then why Should we? Time and Space are expanding, and we must keep our expansion and development "On".

(Contd.)

The Change

The change of the organism is inevitable and gradual, though involuntary, and tangible by the disposition of Space and Time. Humans must be adequately adaptable for sustaining his/her life in an inadvertent competitive surrounding. The comparison and relative strength decide the relative ability and results thereof. The change is defined as the variation of Space and Time concurrently with respect to pre-position of Human. As per physics, there must be an external aid or intervention even for invigorating an infinitesimal movement. The external aid can also be referred to as The Force, which may be exhibited internally or externally or both concomitantly. (In short) Unless the Force acts, things do not move. Change is inevitable, so is Force and its application. The force is a random phenomenon and may be voluntary and involuntary function relates to the Time and Surrounding. Hence, the spread of randomness causes the force to act upon and subsequently instigates "The Change".

The Change is always a spinoff or direct resultant of The Entropy or the disturbances attributing to any point of the Planet and Universe. This means Change comes to an unconcerned and unrelated individual owing to the infinitesimal force application at anywhere on the planet irrespective of location and connection. Do we Welcome Change or Resist! Before answering that another question comes here, "Do we have that opportunity to resist or accept"? Sometimes No and sometimes Yes. The randomness gives Human two chances to opt for and sometimes enacts "The Change" without showering any indications, options and Pre-Triggers.

Life is poised and vulnerable to changes, and the dynamic world is always susceptible to such aspects.

How to Interpret the Change

On prima facie accept The Change, which is natural and voluntary. Welcome The Change! What about the Bad Change, which the Human is certain to encounter in his/her lifetime? Welcome the Change if exhausted by all efforts to prevent it! What about a sweet moment, which you do not want to lose and change obviously? Even here for such case, welcome the Change as there may be upgraded version of current!

Now Going Through the God Point!

- ✦ Why do we change?
- ✦ Why Change knocks us or Vice versa?
- ✦ Why should we accept change even un-wilfully?
- ✦ The Benefits of Change?
- ✦ Can infinitesimal change be neglected?

Why We Change

As stated earlier, The Change is inevitable, random and biological as well. We change physically, Biologically and Mentally so far structural and anatomical aspects are concerned. There are various mediocre aspects of Change and their interpretation. We change by external and internal factors. Our changes are attributable to our voluntary inclination, involuntary action and a state of perplexity. Any of the above factors ensure the changes, maybe desirable or undesirable. Upon such various affecting parameters, we must radically redefine our existence on a real-time basis on this planet for being as more adaptable and flexible every hour it passes. The early Human might have exhibited rugged efforts on resisting change owing to the fear of the future, which they might have never experienced. The change might have downcasted them from the previous state, for which they had a stern disinclination towards change. The Modern Human, in contrast, is favourable to change, conversant and optimistic on future due accruals of thousand years Human thread. Voluntary Change is primarily based on our past experiences, future predictability, Technology and curiosity. In contrary, the involuntary change is resultant of external contributions and an internal mechanism that can be seldom administered. The people incline to change relying on their level of curiosity and learning ability.

However, this may be an obligatory, compulsive, competitive and Sacrifice kind of attempt. On summarisation, people change inevitably and sometimes voluntarily. Change comes to them, and they stride for The Change.

"We Change because we are alive."

"We Change because we have mobility from the cellular level."

"We Change because we have memory and intelligence."

"We Change because we are forwarding with time."

"We Change because we are ageing and certain to die one day."

"We Change because we have emotions, chemical bondage."

"We Change because we lose and gain and are experienced and cynical."

"We Change because of Our Mind, State of Mind and Weird Mind."

"We Change because we do love and hate, fear and dare, hit and miss."

"We Change because we act, reciprocate, react and proact."

"We Change because we can feel, know and experience Change."

"We Change because we Can and Change is acceptable."

Why does Change knock us or vice versa?

Change knocks us, and even we knock change. This is because of the spread of randomness in our lives and everywhere, every step we move. Change knocks us because we are alive, mobile and can sense the Change.

d/t = Change of Position with respect to time

d/t = Change of emotion/ time

d/t = Change of everything/ age

d/t = Change in existence to Nonexistence/Overall age

Converting this equation on a time scale yields as follows:-

Pre-Existence < (d/t) < Existence < (d/t) < Nonexistence

$$\leftrightarrow \leftrightarrow$$

as ***The Change the Change***

Looking at the above order and equations, it can be self-illuminated why "The Change" comes in a broader view.

"Change comes to make us exist on this planet."

"Change comes to enliven our lives and cling between Life and Death."

"We prefer to Change because Change preferred us to exist and decide."

Why should we accept change even without wilfully?

We should accept change because "The Change" accepted us through an incredible Change to our Parents "Parenthood". Our parents might have undergone a series of changes in life, and our existence is a part of their life. Thus, a Change must be accepted wilfully even though sometimes devastating. If somebody encounters unforeseen, unbearable, unacceptable Changes and cannot be averted by any means, the acceptance is the only option as long our existence is progressive and not ceased/threatened. The acceptance can also lead to a state of existence to nonexistence, here the whole great efforts are required to abate such change Fortunately, We Human underwent incremental evolution and finally attained aptness to think and judge towards apropos manoeuvre in life. The change falls under the category of judgement and one must take coherent decision looking into circumstances and outright scenario. "We may accept The Change as long as "It" never had stifled with our existence. The existence is the supreme thing one can have in their life and rest all are Fillers of Life. The fillers of life and change have a proximal connection. The fillers of life comprise a cumulative of infinitesimal changes into an insignificant event. Inadvertently the Insignificant larger change can yield great to greatest changes in life. Hence, Fillers of Life and Change are still crucial factors in one's life. The entity should ensure to differentiate and adept the surrounding by being reactive or proactive.

From the previous explication, it can be comprehended that whether The Change may come through any source or mode of disturbances say involuntarily or voluntarily, inadvertently or astute prediction, the individual has to accept it.

The Change may bring a plethora of happiness or sorrow, curiosity or disinterest, ecstasy or immense pain, leveraged or tormented deeds, the individual has to accept it as long as existence is not threatened. Change is defined as the infinitesimal change of state, behaviour and anything related to the individual with respect to passing time. If The Change is accepted, the next course of action is certain to be controlled to some extent due to vectoring the mind to think about and align the favourable deeds. If some worst thing happened by means of change and that has still been accepted by the individual, then there must be the best thing coming into that individual's life, which could never be scaled to the imagination of a common person, in general. The life altogether is a progressive journey between "2L", and Change is the pivotal puller and ensures "pilotage" of such incremental journey, which may encounter every random, weird, strange, pleasure and divine "Sense of ecstasy" aspects. We must accept Change for the greater good and life continuity with cent per cent willingness to undergo, sustain and leave for the next consequential or subsequential change to occur. There is no question of fear of the future as it is never predictable and can be more beautiful and enchanting and charismatic than the present.

For example, if someone is kept in isolation for 10 years in closed walls, the best Change that could happen is to step out of the door of the isolated room and see the world "as it is" after a decade. Believe it or not, the set freed individual sees the same regular, polluted (environmental) and normal world as "The spectacular, mammoth, extremely joyful and his/her life must be enlivened with basic planet attributes."

In contrast, an individual who has never caged is anticipating a future "which may bring a lot of gratification by fulfilling dreams, milestones and a comfortable life", which may or may not exist at all when scanned all future time frames. Take a case of a freedom fighter who was a strong rebel and had dedicated his/her life but never sees the Anticipating Future of Freedom Day. However, there are certain individuals who never bothered or persuaded for freedom, have seen an enigmatic day of freedom. (Even without any struggle for independence) The perception always relies on behavioural change...

The Benefits of Change

The incredible benefit of The Change is a better and charismatic future than the present. The present and past affect the individual mindset and keep them in Blackface owing to deciding the true future, which should be... That's the reason people quote "Think Positive and Be Positive", which trains the brain for making a sensible decision without being too much decisive about futuristic aspects of life. Being Optimistic is forever good as long as pragmatic virtues are unplacated.

Can Infinitesimal Change Be Neglected?

Even the Big Bang was an infinitesimal those days when the Time and Space were at the nascent stage. However, today, the Big Bang resulted in infinite content of matter of the infinite space... So the infinitesimal vicissitudes should not be neglected; instead slight attention should be paid so as to adjugate the consequential prospects, which is often misinterpreted. In the process of being too much judicious for infinitesimal or minuscule aspects, an individual may lose conscious and coherent decisive strength. Therefore, attention should be differentiated with intricate future predictions. In previous chapters, similar contents and pertaining description are elaborated. Readers may refer to the first chapter of randomness, fillers and Entropy aspects for a smaller change.

"The End is an End"

The God Point states that eternal milestones of life are nonexistence. The nonexistence is the absolute ultima of life and is a pure state, which is the most benign state and dormant until there is a birth or existence by any means. The life commencement is a miraculous transcendence from nonexistence to existence, indeed an inadvertent process of fantasy. The life instigation has equal significance as Life destruction. Hence, "The beginning is a beginning, and the end is an End." All those happen in between are "The Fillers of Life". When someone knew that Life would end up someday, what is the necessity of such struggle he/she encounters voluntarily or involuntarily? This is because of Survival Interest, Pain bearing ability, never fulfilling gratification and never-ceasing Curiosity, all of them enliven a life … until the body is fit to retain the Life Energy qualitatively and quantitatively. Things to ensure in the current Life, in fact, The Only Life, are two trivial concepts in Life:

1. The Beginning: Why, How, When, Outcome and Caution
2. The End

The beginning: The beginning is the beginning, so always be committed to starting with a positive mind. Eventually, the good beginning seldom disappoints the way of outcome and results to occur. However, before the beginning, one must stick to basic ideologies. Upon exception, simple things will be intricate yielding greater variability of the resultants.

Why we Begin:

Basic necessity, Curiosity, Expectations, Inadvertently, Dependency, Interdependence, for being independent or self-reliant, Help, Protection.

- How to Begin:
  - Be happy and Be optimistic
  - A precise understanding of results or outcomes and a rough idea of approach
  - Check expectations
  - Impeccable dedication—train to brain
  - Erase former memories
  - Refer precedence to an acceptable level
  - Keep all types of relation in isolating path; ensure not to cross the emotional, behavioural paths. For example keep house, home, office and leisure aspects in parallel but those should never be convergent or interceptive.
  - When to Begin:
  - People often say right now, to some desirable extent that is okay, however not always.

- Begin when situation demands and your brain pushes you more than your heart.
- Begin before "A triggering event", which instigates an inadvertent or undesirable beginning for you.
- In rare cases, End before beginning.

✦ Outcomes:
- The beginning has an outcome before the overall outcome.
- The outcome of the beginning is the Initial Perception and feelings of sense within individuals.
- The initial perception should neither be neglected nor be kept under significance. Ensure the initial perception to be aligned with a pragmatic world coupled with optimism. All that can happen by "Train the Brain" and "Don't train hormones or body chemicals". The hormones and chemicals always are counterintuitive due to propensity to impede the changes within, by surrounding and a great fear owes to an elevated level of pre-existing protectiveness in *Homo sapiens*.

✦ Cautions before Beginning:

✦ Being Optimistic is utterly crucial but be in the purview of pragmatic aspects.

✦ Expectation is the true trigger for the commencement of any activity but it never yields equated results. Truly speaking, expectations are random and will never match each other (refer at the billionth level of Space/Time emulation of event-to-event, outcome to outcome).

- ✦ Time is the best guide sometimes, so have patience before beginning and until the right time to take a commencement call.

The End

The end can be deduced into such steps as the beginning but let us keep it simple to the extent of a dot.

"Life is not coming back, and we have to live to fullest else we will certainly miss the opportunity. If there is a meter to measure happiness, gratification, etc., we the planet earthman are terribly scoring low and remain lower percentile quarter. If another living planet exists, those individuals living on it are certain to remain happier and more satisfied than we are. This is because, our planet, habitat and brain development made us more competitive, struggle oriented and denounced of mental peace by ourselves." This cannot be corrected at least by the span of thousand years, but there is an instant solution.

The solution is the

"Instant Happiness and Satisfactory life"

To attain this, we have to ensure

A Sound Health: Yes Health and the Human body organs are most precious. Ensure to have a good health and avoid damages to Health by your own acts, habits and irregularities.

Train the Brain: Yes. Please train your brain and body chemicals to remain cent per cent happy and satisfied as always. Being competitive is good; protective is good but anything in excess leads to devastation. Train the brain for creative delusive-happiness always, someday, the brain will absorb Absolute Happiness in reality even in a chaotic turbulent unbearable scenario, as far as the mental state of mind is concerned.

Just Say I Am Happy Biologically: My possessions and relations concerned with this planet and Planet mate are absolutely temporary and train the brain not to release connecting and affecting chemicals. The brain can take you to ecstasy or Hell; all depends on you, yes on the individual behaviour.

Stay away from Surrounding Mentally: Discern good and bad before being entrapped in complexity.

Do Risk Mitigation and Manage Personal Risks

A. Money and Security Fear: This is important and inevitable. This cannot be assured to zero but can be managed like a Risk Management by better understanding terms with connected people and close circle. Say Spouse and Children and Parents.

B. Medical Risk assessment and Security must be planned. Emergency Services and actions should always be in the vicinity, planned well before. Being positive is good under pragmatic purviews like medical emergencies and actions. Recurrently Train the Brain in every step and days in Life.

"Train the brain Everyday like we follow our daily routine".

C. Do not be a slave of your intricate trap. Come out of the trap, do not work for money and obligation and find some gratified jobs you are best with and capable of. This is a conceptual hypothesis, but possible as long as your skill level is growing incrementally. Job stress is for an employee and not for the chairperson of the company. Think in perspective of an Owner, you will feel pity for your previous thinking.

D. Don't let others control you beyond your will. If it happens, change and diverge from the place, propel. You cannot change them but can change you and your belongings. Ensure risk mitigation well before terminating earning prospects in life. You can be happy and adjust yourself without desirable wealth, but your dependents cannot. So train them as well. However, that is a quasi-critical idea but possible with relentless preservice on a long-term convincing and understanding.

E. Think before you think about others by inter-switching the brain positions. This will help to reduce major confrontations, arguments and enrich good social life.

F. Be Introvert or Extrovert that really does not matter but be a Human and ensure humanity is exhibited in all activities that are being executed. Help Others.

G. Ensure being in Higher Potential, By Thinking, not on earning or conducive living style. Keep your thoughts elevated far from reach of common diverging aspects. This will always enable you to make decisions holistically and not situation and person or evidence-based.

H. Be Creative and Innovative; create anything as much as you can. This will facilitate you of doing an unconventional, non-redundant and engaging jobs or endeavours, which you targeted or inadvertently executed.

I. Be Pragmatic and Do not panic for death-related aspects. Even if an individual who has a time-bound dwell under this Sun and on the Planet Earth expects a long term as if time will retain for him/her. All the people leave this planet with Space and Time variations; in fact, the nonexistence is absoluteness of everything that one has ever thought of. Remember the best thing that could happen to any individual in this planet is

"If He/she has never existed on this planet". Yes, Unquestionably. Nonexistence is the best thing that could happen. Thus,

Inexperience is better than experience, as long as experiences are not experienced and compared. Not getting a chance to experience and having an ability to experience, eternally, are indeed best experiences ever, State of being in heaven ever or Being in ecstasy ever without progression of Space and Time is ultimacy one can be at.

That's why $E = MC^2$ is "The Heaven"

Heaven = MC^2 (Refer energy for Liberation = MC^2, in previous subchapter).

The Heaven is Indeed the incredible absoluteness and a magnificent state of status quo and Immaculately Heavenly experiences of nothing.

"Nothing is everything that can never be lost, enervate or consumed; Nothing can't lose its ingenuity." Therefore, I prefer the end is superior than the beginning and is an immaculate end, which comes eventually in everyone's life.

Penultimately:

The end is the end. "Be utterly Happy every minute you live before experiencing nothingness." *"You must be happy because you can experience happiness."* The end will never be altered, and the end can never instigate a recurrence as well. Human life is as equivalent to the life of any cell of the body, Just Space and Time matter. The time enervation matters. The best what we can contribute to this planet is to preserve its habitation and enable life sustainability factors. Eventually, a planet mate will help the succeeding generations thereby enriching their lives, which are relatively better but not the ultimacy.

Nothing can be ultimate than nothingness, which encapsulates everything within itself in the status quo with respect to Space/Time, Experiencing ability and Merging into a dot of pre-existence or nonexistence.

As the end is the end, ensure being healthy, do not fall prey to other contemplations, be yourself, give time to you and connected people. The planet mate so-called intellectual is not that intellectual to alter space/time; never listen unless your inner conscience permits. Experience and celebrate this world as a rental planet, a rental body you pose and rental life experiences, which are pervasively diffusing. Count on best experiences in life; at least ensure there are not too many bad experiences in life by pseudo-comparisons and possessions. Celebrate your Human life, it may be only a one-time opportunity in the Space-Time Basket of Universe.

"Nothing is better than everything put together."

"Inexperience is better as long as we don't have the ability to experience."

"The End is an Eternal End."

www.ingramcontent.com/pod-product-compliance
Lightning Source LLC
Chambersburg PA
CBHW030940180526
45163CB00002B/642